T0272601

FEEDING
THE
MACHINE

FEEDING THE MACHINE

THE HIDDEN HUMAN LABOR POWERING A.I.

JAMES MULDOON, MARK GRAHAM, and CALLUM CANT

BLOOMSBURY PUBLISHING

NEW YORK • LONDON • OXFORD • NEW DELHI • SYDNEY

BLOOMSBURY PUBLISHING
Bloomsbury Publishing Inc.
1385 Broadway, New York, NY 10018, USA

BLOOMSBURY, BLOOMSBURY PUBLISHING, and the Diana logo
are trademarks of Bloomsbury Publishing Plc

First published in 2024 in Great Britain by Canongate Books Ltd
First published in the United States in 2024 by Bloomsbury Publishing

Copyright © James Muldoon, Mark Graham, and Callum Cant, 2024

ISBN: HB: 978-1-63973-496-2; eBook: 978-1-63973-497-9

Library of Congress Cataloging-in-Publication Data is available.

2 4 6 8 10 9 7 5 3 1

Typeset by Palimpsest Book Production Ltd, Falkirk, Stirlingshire, UK
Printed and bound in the U.S.A.

To find out more about our authors and books visit
www.bloomsbury.com and sign up for our newsletters.

Bloomsbury books may be purchased for business or promotional use.
For information on bulk purchases please contact Macmillan Corporate
and Premium Sales Department at specialmarkets@macmillan.com.

Contents

Authors' Note

The names of individuals referred to by first name only in this book have been anonymized, and certain identifying details have been fictionalized. Any resemblance between a fictionalized individual and a real person with the same name is strictly coincidental.

FEEDING
THE
MACHINE

The Extraction Machine

Mercy craned forward, took a deep breath and loaded another task on her computer. One after another, disturbing images and videos appeared on her screen. As a Meta content moderator working at an outsourced office in Nairobi, Mercy was expected to action one 'ticket' every fifty-five seconds during her ten-hour shift. This particular video was of a fatal car crash. Someone had filmed the scene and uploaded it to Facebook, where it had been flagged by a user. Mercy's job was to determine whether it had breached any of the company's guidelines that prohibit particularly violent or graphic content. She looked closer at the video as the person filming zoomed in on the crash. She began to recognise one of the faces on the screen just before it snapped into focus: the victim was her grandfather.

Mercy pushed her chair back and ran towards the exit, past rows of colleagues who looked on in concern. She was crying. Outside, she started calling relatives. There was disbelief – nobody else had heard the news yet. Her supervisor came out to comfort her, but also to remind her that she would need to return to her desk if she wanted to make her targets for the day. She could have a day off tomorrow in light of the incident – but, given that she was already at work, he pointed out, she may as well finish her shift.

New tickets appeared on the screen: her grandfather again, the same crash over and over. Not only the same video shared by others, but new videos from different angles. Pictures of the car; pictures of the dead; descriptions of the scene. She began to recognise everything now. Her neighbourhood, around sunset, only a couple of hours ago – a familiar street she had walked along many times. Four people had died. Her shift seemed endless.

We spoke with dozens of workers just like Mercy at three data annotation and content moderation centres run by one company across

Kenya and Uganda. Content moderators are the workers who trawl, manually, through social media posts to remove toxic content and flag violations of the company's policies. Data annotators label data with relevant tags to make it legible for use by computer algorithms. We could consider both of these types of work 'data work', which encompasses different types of behind-the-scenes labour that makes our digital lives possible. Mercy's story was a particularly upsetting case, but by no means extraordinary. The demands of the job are intense. 'Physically you are tired, mentally you are tired, you are like a walking zombie,' noted one data worker who had migrated from Nigeria for the job. Shifts are long and workers are expected to meet stringent performance targets based on their speed and accuracy. Mercy's job also requires close attention – content moderators can't just zone out, because they have to correctly tag videos according to strict criteria. Videos need to be examined to find the highest violation as defined by Meta's policies. Violence and incitement, for instance, are a higher violation than simple bullying and harassment – so it isn't enough to identify a single violation and then stop. You have to watch the whole thing, in case it gets worse.

'The most disturbing thing was not just the violence,' another moderator told us, 'it was the sexually explicit and disturbing content.' Moderators witness suicides, torture and rape 'almost every day', commented the same moderator; 'you normalise things that are just not normal.' Workers in these moderation centres are continually bombarded with graphic images and videos and given no time to process what they are witnessing. They're expected to action between 500 and 1,000 tickets a day. Many reported never feeling the same again: the job had made an indelible mark on their lives. The consequences can be devastating. 'Most of us are damaged psychologically, some have attempted suicide . . . some of our spouses have left us and we can't get them back,' commented one moderator who had been let go by the company.

'The company policies were even more strenuous than the job itself,' remarked another. Workers at one of the content moderation centres we visited were left crying and shaking after witnessing beheading videos, and were told by management that at some point during the week they could have a thirty-minute break to see a 'wellness counsellor' – a colleague who had no formal training as a psychologist. Workers who ran away from their desks in response to what they'd seen were told they had committed a violation of the company's policy because they hadn't remembered to enter the right code on their computer indicating they were either 'idle' or on a 'bathroom break' – meaning their productivity scores could be marked down accordingly. The stories were endless: 'I collapsed in the office', 'I went into a severe depression', 'I had to go to hospital', 'they had no concern for our wellbeing'. Workers told us that management was understood to monitor hospital records to verify whether an employee had taken a legitimate sick day – but never to wish them better, or out of genuine concern for their health.

Job security at this particular company is minimal – the majority of workers we interviewed were on rolling one- or three-month contracts, which could disappear as soon as the client's work was complete. They worked in rows of up to a hundred on production floors in a darkened building, part of a giant business park on the outskirts of Nairobi. Their employer was a client of Meta's, a prominent business process outsourcing (BPO) company with headquarters in San Francisco and delivery centres in East Africa where insecure and low-income work could be distributed to local employees of the firm. Many of the workers, like Mercy herself, had once lived in the nearby Kibera slum – the largest urban slum in Africa – and were hired under the premise that the company was helping disadvantaged workers into formal employment. The reality is that many of these workers are too terrified to question management for fear of losing their jobs. Workers reported that those who complain are told to shut up and reminded that they could easily be replaced.

While many of the moderators we spoke to were Kenyan, some had migrated from other African countries to work for the BPO and assist Meta moderate other African languages. A number of these workers spoke about being identifiable on the street as foreigners, which added to their sense of being vulnerable to harassment and abuse from the Kenyan police. Police harassment wasn't the only danger they faced. One woman we interviewed described how members of a 'liberation front' in a neighbouring African country found names and pictures of Meta moderators and posted them online with menacing threats because they disagreed with moderation decisions that had been made. These workers were terrified, of course, and went to the BPO with the pictures. The company informed them they would see about enhancing security at the production facilities; aside from that, they said, there was nothing else they could do – the workers should just 'stay safe'.

Most of us can hope never to experience the inhumane working conditions endured by Mercy and her colleagues. But datawork like this is performed by millions of workers in different circumstances and locations around the world. At this particular centre, some of the working conditions changed after our fieldwork was conducted, which we address in more detail in Chapter 8. But large companies like Meta tend to have multiple outsourced providers of moderation services who compete for the most profitable contracts from the company. This data work is essential for the functioning of the everyday products and services we use – from social media apps to chatbots and new automated technologies. It's a precondition for their very existence – if it wasn't for content moderators continually scanning posts in the background, social networks would be immediately flooded with violent and explicit material.[1] Without data annotators creating datasets that can teach AI the difference between a traffic light and a street sign, autonomous vehicles would not be allowed on our roads. And without workers training machine learning algorithms, we would not have AI tools such as ChatGPT.

Artificial intelligence can be broadly understood as a machine-based system that processes data in order to generate outputs such as decisions, predictions and recommendations.[2] It can refer to anything from autofill in emails to targeted weapons systems in drone warfare. The reality is it's more of a marketing concept, or an umbrella term under which very different technologies can be grouped. This includes computer vision, pattern recognition and natural language processing (that is, the processing of everyday speech and text). It's an amorphous idea that can evoke the wonders of post-human intelligence but also herald the dangers of an AI-triggered extinction event. AI stands in for many different ideas in public debate: economic growth, scientific achievement and advanced capabilities for some, at the same time as automated job losses, biased decision-making and tech evangelism for others. Its meaning has also evolved over the years as it is constantly redefined to capture the latest wave of tech development.

Most recently, this has centred upon the systems that power chatbots: large language models or LLMs. LLMs are trained on enormous datasets containing vast amounts of text data usually scraped from the Internet. Large language models such as ChatGPT are called large because of the size of their datasets (hundreds of billions of gigabytes of data), but also because of the number of parameters that have been used to train them (about 1.76 trillion parameters for ChatGPT-4). Parameters are the variables that drive the performance of the system and can be fine-tuned during training to determine how a model will detect patterns in its data, which influences how well it will perform on new data.

Today, we are in the middle of a hype cycle in which companies are racing to integrate AI tools into a variety of products, transforming everything from logistics to manufacturing and healthcare. AI technologies can be used to diagnose illnesses, design more efficient supply chains and automate the movement of goods. The global AI market was worth over $200 billion in 2023, and is expected to grow 20 per cent each year to nearly $2 trillion by 2030.[3] The development of AI tends to be secretive and opaque; there are no exact numbers of how many workers participate globally in the industry, but the figure is in the millions and, if trends continue at their current rate, their number will expand dramatically.

By using AI products we are directly inserting ourselves into the lives of these workers dispersed across the globe. We are connected whether we like it or not. Just as drinking a cup of coffee implicates the coffee drinker in a global production network from bean to cup, we should all understand how using a search engine, a chatbot – or even something as simple as a smart robot vacuum – sets in motion global flows of data and capital that connect workers, organisations and consumers in every corner of the planet. Many tech companies therefore do what they can to hide the reality of how their products are actually made. They present a vision of shining, sleek, autonomous machines – computers searching through large quantities of data, teaching themselves as they go – rather than the reality of the poorly paid and gruelling human labour that both trains them and is managed by them. But actions we take as consumers, activists and citizens can have a real impact on the conditions of these workers and help them in their struggle for decent work. These workers are at the forefront of these technological changes, but AI-enabled surveillance and productivity tools are coming for many workers, even those who might consider themselves immune from such encroachment into their working lives. The first step towards taking action is understanding how AI is produced and the different systems that are at play. This allows us to see how AI helps concentrate power, wealth and the very ability to shape the future into the hands of a select few.

This book tells the stories of the people whose labour makes AI possible and describes the systems of power that maintain global inequalities in access to capital, networks and work opportunities. It exposes the hidden human workforce that contributes to AI and reveals how this essential work is often deliberately concealed. Based on over 200 interviews with data annotators, content moderators, machine learning engineers, AI ethicists, warehouse workers, labour organisers and industry figures, *Feeding the Machine* sheds light on the hidden world of AI production and its overlooked digital workers. These workers find themselves in a variety of positions – from the ultra-precarious and low-income data annotators working in poor conditions without minimum employment protections, to the well-remunerated machine learning engineers with high salaries in the headquarters of global tech companies. By following the trail of money that connects these AI production networks, we can begin to uncover the deep history upon which AI is built, and reveal the colonial legacies that still shape it today.

Drawing on over a decade of research experience, each chapter of this book takes the reader to a different location in which workers contribute to the production of AI. In the following pages, we will meet seven people who play a key role in this process. These figures are all real people we have interviewed as part of our research – every story in the book is true – but in some cases we have taken careful steps to disguise them, on occasion eliding more than one person from our interviews. The seven figures are 'the annotator', 'the engineer', 'the technician', 'the artist', 'the operator', 'the investor', and 'the organiser'. By showing their work process, social context and daily lives, each chapter provides a glimpse into the human element of AI production and its impact on workers. Among these figures, we will meet a Ugandan data annotator trapped in a mind-numbing and tedious job, but whose opportunities are so limited that she feels she has nowhere else to go. We will meet an Irish voice actor who realises her voice has been synthesised without her knowledge by a machine learning algorithm that could eventually replace her. We will also learn of a Kenyan political activist who organises with his fellow workers and struggles to change an unjust system that only serves the rich and powerful of the digital economy. The overarching narrative reveals how the work of each of these seven figures is connected and how the actions of one can have drastic consequences for the lives of the others.

The purview of this book is necessarily partial and incomplete; it's not possible to cover every single aspect of how AI is produced or the diverse ways in which it is used in a single book. In our chapters on individual workers who power AI we don't, for example, devote an entire chapter to the miners who excavate the critical minerals needed for certain tech products, or the labourers who assemble these products in factories around the world – often under appalling conditions. Many other figures could have easily filled the pages of this book. Instead, we attempt to offer a series of snapshots that present different perspectives on AI, ones that move beyond the narrow picture of Silicon Valley. In doing so, the book will take us on a journey through Kenya, Uganda, Ireland, Iceland, the UK and the US.

Artificial intelligence is often conceived of as a mirror of human intelligence, an attempt to 'solve intelligence' by reproducing the processes that occur within a human mind. But from the perspective we develop in this book, AI is an 'extraction machine'. When we engage with AI products as consumers we only see one surface of the machine and the outputs it produces. But beneath this polished exterior lies a complex network of components and relationships necessary to power it. The extraction machine draws in critical inputs of capital, power, natural resources, human labour, data and collective intelligence and transforms these into statistical predictions, which AI companies, in turn, transform into profits. To understand AI as a machine is to unmask its pretensions to objectivity and neutrality. Every machine has a history. They are built by people within a particular time to perform a specific task. AI is embedded within existing political and economic systems and when it classifies, discriminates and makes predictions it does so in the service of those who created it. AI is an expression of the interests of the wealthy and powerful who use it to further entrench their position. It reinforces their power while at the same time embedding existing social biases in new digital forms of discrimination.

Corporate narratives of AI emphasise its intelligence and convenience, often obscuring the material reality of its infrastructure and the human labour needed for it to function.[4] In the public imagination, AI is associated with images of glowing brains, neural networks and weightless clouds, as if AI itself simply floated through the ether. We tend not to picture the reality of the constant heat and white noise of whirring servers loaded into heavy racks at energy-intensive data centres, nor the tentacle-like undersea cables that carry AI training data across the globe. AI has a material body and exists only through new chips, servers and cables being manufactured and added to the machine. And just like a physical body, AI's material structure needs constant nourishment, through electricity to power its operations and water to cool its servers. Every time we ask ChatGPT a question or use an Internet search engine, the machine lives and breathes through this digital infrastructure.

We also tend to forget that behind the seemingly automated processes of AI often lies the disguised labour of human workers forced to compensate for the limitations of technology.[5] AI relies on human workers to perform a wide variety of tasks, from annotating datasets to verifying its outputs and tuning its parameters. When AI breaks down or does not function properly, human workers are there to step in and assist algorithms in completing the work. When Siri does not recognise a voice command or when facial recognition software fails to verify a person's identity, these cases are often sent to human workers to establish what went wrong and how the algorithm could be improved. The original 'Mechanical Turk' (after which Amazon's crowdsourcing marketplace was named) was a fraudulent chess-playing machine with a life-sized figure of a man carved from wood that toured Europe in the late eighteenth century.[6] The inventor of the machine, Wolfgang von Kempelen, claimed it could play chess automatically of its own accord, but hidden inside the box was a human chess master who operated it through a series of levers and mirrors. The idea that today's AI works autonomously relies on a similar illusion. Sophisticated software functions only through thousands of hours of low-paid and menial labour – workers are forced to work like robots in the hopes that AI will become more like a human.

The extraction machine not only requires physical resources and labour to function, but lives off the human intelligence contained in its training datasets. AI captures the knowledge of human beings and encodes it into automatic processes through machine learning models. It is fundamentally derivative of its training data, through which it learns to undertake a diverse range of activities: from driving a car to recognising objects and producing natural language, and it relies on a project of collecting the history of human knowledge in enormous datasets consisting of billions of data points. The systems trained by these datasets can often perform at superhuman levels, and while many of these datasets are in the public domain, others contain copyrighted works taken without their authors' consent. AI companies have undertaken a privatisation of collective intelligence by enclosing these datasets and using proprietary software to create new outputs based on a manipulation of that data. The extraction machine requires these intellectual resources as much as the material ones.

But we don't just call this system an extraction machine because it's produced through the plunder of resources, human labour and our collective intelligence. When AI systems get put into action, particularly in the workplace, they also facilitate further processes of extraction. Primarily, this is the extraction of effort from workers, who are forced to work harder and faster by AI management systems that centralise knowledge of the labour process and reduce the level of skill required to do a job by routinising and simplifying it. This intensification of work extracts more value from the labour of workers for the benefit of employers. For many of us, this will be the mechanism through which we are most exposed to the damage caused by the extraction machine. We might not become content moderators any time soon, but the same machine that entraps Mercy affects our jobs too.

Let us take one example of how this extraction machine operates. A German car manufacturer operates as a lead firm in a global production network, coordinating a range of supplier firms scattered around the world. While each of these companies plays a crucial part in developing the final product, it is ultimately the car company that coordinates the network. No other single firm has the same level of oversight or knowledge about how the entire system functions. In this example, the company decides to produce cars with level 3 automated driving – known as 'conditional' automation, where a driver can take their eyes off the road in certain conditions. This sets in motion a chain of effects that involves workers and organisations from across the globe.

The directors of this large company have been encouraged by institutional shareholders to explore the autonomous driving space following the rise in the value of a competitor's shares when it launched its own programme. The company compiled a list of publicly available datasets and selected what it needed, then purchased private datasets with thousands of hours of footage annotated with labels from hundreds of object categories (traffic lights, pedestrians, other motorists etc.). After developing a model at its AI lab, machine learning engineers at the company realised that there were several edge cases (rare events or scenarios) for which they would need new data annotated to further train the model. Datasets consisting of cars driving in different conditions have to be manually annotated by thousands of annotators. The company engages three different annotation providers in the Philippines, Kenya and India to perform these tasks. Once the machine learning engineers review the annotated data and attempt to fine-tune the model, several batches of work are sent back to the data annotation companies to redo.

The lab also rents computational – or compute – resources (the processing power, memory and storage needed to run programs) from Amazon Web Services, and although the German team are unhappy about sending their sensitive data to an external provider, the company had to organise to rent these specialised AI services months in advance. Amazon is one of the few providers that can offer the speed and scale the company needs.

Once the model is developed and passes several stages of safety testing, it can be marketed to customers and put out on the roads. Many of the early stage customers for this type of technology are likely to be large logistics firms looking to cut costs and gain a competitive advantage. Trucks that are capable of self-driving between hubs on motorways, for example, may allow employers to reorganise drivers' jobs around short-haul deliveries from hubs to end customers, increasing productivity and managerial control. From start to finish, the production and deployment of AI systems extracts labour, resources, intelligence and value.

Studies of the broader social context in which AI operates are increasingly important, since we are entering a new era of tech development. The 2010s were characterised by the growth and then dominance of a handful of digital gatekeepers that amassed billions of users on their platforms, became trillion-dollar companies and leveraged their position to exercise unparalleled political and economic power. The rise of AI has led to major shifts in the internal dynamics of the tech sector, which has profound consequences for the global economy. The platform era that lasted from the mid-2000s to 2022 has now given way to an era of AI. Following the launch of ChatGPT and new partnerships between Big Tech and AI companies, both investment strategies and business models are driven by a new coalescence of forces around AI.

The era of AI has given rise to a new configuration of major players that overlaps with but is distinct from the platform era. In place of the leading Big Tech firms of the 2010s, a group of companies we call 'Big AI' has emerged, and are central organisations of this new era. This group of companies includes legacy Big Tech firms such as Amazon, Alphabet, Microsoft and Meta and also includes AI startups and chip designers like OpenAI, Anthropic, Cohere and Nvidia. If attention was turned to Chinese companies, which are the next most significant set of actors in the era of AI, we could also include Alibaba, Huawei, Tencent and Baidu. Although the precise membership of this group is likely to shift, Big AI consists of companies that understand AI as a commercial product that should be kept as a closely guarded secret, and used to make profits for private companies. Many of these companies seek to limit knowledge about how their AI models are trained, and develop them in ways that increase their competitive advantage in the sector. Following the public release of ChatGPT, a series of new strategic collaborations was announced between legacy tech firms and AI startups. Microsoft invested $10 billion in OpenAI; Google invested $2 billion in Anthropic; Amazon invested $4 billion in Anthropic; Meta has partnered with both Microsoft and AI startup Hugging Face; Microsoft developed a new AI unit from Inflection staff members; while Nvidia is now a two trillion-dollar company that supplies 95 per cent of the graphics processing unit (GPU) market for machine learning.[7]

The dominance of social media and advertising platforms during the platform era was partially based on 'network effects': the more users a platform had, the more efficient and valuable its service became and the more profitable for its owners. Large quantities of user data provided platform owners with greater insight into this digital world and the ability to better extract value through fees or advertising revenue. In the era of AI, ownership over software still matters, but the underlying hardware has grown in importance. Early platform companies were lean: Airbnb did not own any houses and Uber did not own any cars. They were selling X-as-a-service and relied on networks of users to make it all happen. Big AI benefits from what we call 'infrastructural power': ownership of AI infrastructure – the computational power and storage needed to train large foundation models. This occurs through their control of large data centres, undersea fibre-optic cables, and AI chips used to train their models. Just three companies own over half of the world's largest data centres, while only a select few can provide access to the hardware needed to train cutting-edge AI models. This infrastructural power also exercises a profound pull on AI talent, because the best people in the industry want to work at the leading organisations where they can do state-of-the-art work on the development of AI. Rather than AI opening the doors to more innovation and diversity, we may be witnessing the further consolidation of wealth and power as new players join more established firms.[8]

One consequence of this infrastructural power is a change in the nature of funding models and the degree of independence for new startups. AI companies do not just require a few million to get started – they need hundreds of millions in capital and access to a cloud platform to train foundation models. This means AI startups require strategic partnerships with existing cloud providers that often buy a minority stake in the company. Large tech companies are also in a perfect position to provide billions in funding to new startups because they tend to have large cash reserves. The first generation of platforms received funding from venture capital (VC), but the original founders maintained significant unilateral control over their businesses. As a result, many of these platforms turned into gigantic empires ruled by a single billionaire founder. This is unlikely to occur in the era of AI, because any new empires will have to cooperate or merge with existing mega-corporations. The struggle to successfully commercialise AI products will likely create a multi-polar tech sphere in which legacy tech companies seek to partner with the most successful of the younger startups to form new coalitions to outcompete their rivals.

Perhaps the most interesting and yet unknown shift will be the change in business models adopted by the leading AI companies. The most infamous model of the platform era is the advertising platform, epitomised by Facebook and Google and critiqued in Shoshana Zuboff's concept of 'surveillance capitalism'.[9] The surveillance capitalism model of providing a free digital service in exchange for selling targeted advertising to users is of course not the only business model of platforms – Amazon has a monopolistic marketplace, Uber and Airbnb charge transaction fees, and Netflix and Spotify have a subscription model – but it is a defining characteristic of the era. Yet it's unclear whether the surveillance paradigm still has as much purchase on the new generation of AI companies; the 'surveillance' aspect of Zuboff's theory was always just a business model rather than a new modality of capitalism. Tech companies are more than happy to earn revenue through other methods if they will be just as profitable. What precisely will emerge is yet to be determined, but we can see early signs of AI companies generating revenue through licensing, subscriptions, integrating AI into existing services and renting out 'AI-as-a-service'.

Finally, the era of AI will unfold against the backdrop of a more extreme and divided geopolitical context shaped by the climate crisis, resource insecurity and tensions between the United States and China, all of which will profoundly shape the development of AI. Digital platforms have always been connected to forms of security, surveillance and border technology. The difference between AI and the first generation of social media apps and online marketplaces is the degree to which governments perceive it to be a direct tool to enhance their military and economic power. We are returning to a much more antagonistic relationship between the United States and China where – much like in the Cold War – technology is seen as a marker of civilisational achievement, a method of developing advanced weaponry and a means to gain an economic competitive advantage.

This geopolitical rivalry must also contend with the rising importance of sustainability, a concern that is taken seriously, at least on paper, by all of the major tech companies. In the platform era, the environmental costs of digital infrastructure were not as widely reported on as they are today. Sustainability concerns also shape different countries' access to the critical minerals needed to develop advanced AI chips, and the role of specific geographic regions in mining and processing them. All these factors will influence the types of AI that are developed and how they will be deployed in an increasingly precarious world.

There are also profound continuities between the current era and past decades of tech development. If anything, current trends can best be described as the growth of tech companies' ambitions for global dominance and the expansion of their empires deeper into the social fabric of our lives and the halls of political power. AI accelerates these trends and enriches those who have already benefited from the growing concentration of power in the hands of American tech billionaires. For those at the bottom of the pile, the pickings will be slim indeed.

If countries in the Global South had little say in how digital surveillance platforms were built or deployed in their neighbourhoods, they will have even less input into the development of AI – a technology that is shrouded in mystery and requires enormous resources and computational power. The extraction machine will require much from these countries – but as raw material, to be fed as inputs into its voracious mechanical core.

How did we come to write this book on the human labour powering artificial intelligence? We are three researchers on technology and politics who work together at the Oxford Internet Institute, University of Oxford, as part of Fairwork for AI, a research project aimed at better understanding how workers contribute to building AI systems, and ultimately at improving their working conditions. Mark is a professor and the director of the project, and Callum and James were research associates and are now lecturers at the University of Essex. Our diverse backgrounds stretch across sociology, political science, geography, history, law and philosophy, all of which have been brought to bear on both Fairwork's research and, subsequently, this book.

When commentators engage in abstract debates about what AI might become decades into the future, including its possible social harms, it can be easy to forget the importance of the humans building it *right here in the present*. Speculative thinking about the existential threats of Terminator-like systems can draw much-needed attention away from an analysis and critique of the powerful interests that lie behind AI today. When focusing on the present, a number of studies have already laid bare the real threat of biased and discriminatory outcomes of AI systems. There is, however, less analysis of how such systems of exclusion permeate both the production process, negatively impacting women, minorities and workers from the Global South, and the deployment of AI systems in the workplace and broader society.

This book combines two elements for the first time: a deep economic and political analysis of the systems of labour that produce artificial intelligence, and a rich ethnographic account of workers' lives and how their work contributes to broader production networks. It is not simply a survey of data annotators and other workers, featuring stories from across the world; it is a critique and exposé of the systems that relentlessly maintain global inequalities in the digital economy.

In *Feeding the Machine*, we draw a line from the technological development of current AI systems back to earlier forms of labour discipline used in industrial production. We argue that the practices through which AI is produced are not new. In fact, they closely resemble previous industrial formations of control and exploitation of labour. Our book connects the precarious conditions of AI workers today to longer histories of gendered and racialised exploitation – on the plantation, in the factory, and in the valleys of California. We build on existing conversations about the development of technologies of control employed by tech companies to discipline and manage their workers. To understand these connections, we must trace the history of work outsourced to cheaper and more disciplined labour in emerging markets during the 1990s, and the new technologies of control that accompanied it.

The emergence of the networks through which AI is produced echoes colonial histories of extraction and exploitation through plunder and lopsided trade agreements. Colonialism is generally considered as the territorial appropriation by an empire of the natural environment and human labour of a colony. But Latin American decolonial scholars remind us that colonialism's effects still endure post-liberation through what they call a structure of 'coloniality' – a system of power that defines culture, labour and knowledge production based on older colonial hierarchies.[10] To properly understand AI, we have to view its production through the legacy of colonialism.[11]

Coloniality is part of the structuring logic of artificial intelligence – both in how it is produced and how it operates. AI is produced through an international division of digital labour in which tasks are distributed across a global workforce, with the most stable, well-paid and desirable jobs located in key cities in the US, and the most precarious, low-paid and dangerous work exported to workers in peripheral locations in the Global South. Critical minerals required for AI and other technologies are mined and processed in locations across the Global South and transported to special assembly zones to be turned into technology products such as the advanced AI chips required for large language models. These practices continue well-worn colonial patterns of Western countries leveraging their economic dominance and growing rich off extracting minerals and labour from peripheral territories. Outputs from generative AI also reinforce old colonial hierarchies, since much of the AI datasets and common benchmarks on which these models are trained privilege Western forms of knowledge and can reproduce damaging stereotypes and display biases against minority groups misrepresented or distorted in the data.

Nowhere is the link between the colonial past and the present day more apparent than in Mercy's home of Kibera. The district, part of the Nairobi conurbation, possesses the dubious distinction of being the largest urban slum in Africa. Because of the informal nature of the settlement, nobody quite knows how many people live there. Estimates range from 200,000 to a million. To make a living, most residents find jobs in the diverse informal economy: working in little shops and selling second-hand clothes, homewares, mobile phones, or food. A lot of those shops line a single-track railway that cuts Kibera in half. Trains pass through a few times a day. But, at other times, the track makes a convenient pathway for residents because it's elevated from the dense, congested alleyways of the slum.

When we visited Mercy at her home in Kibera, we spent some time walking along those train tracks. As we stepped over old wooden railway ties that were now almost entirely submerged under compacted soil, we looked out over the slum's makeshift housing. Buildings were made from dried mud, occasional slabs of concrete, wooden poles, and had corrugated iron sheets for roofs. Few residents have indoor plumbing, and the smell of open sewers is often overpowering. Life expectancy is low, illiteracy is high, and residents struggle to break from the cycle of poverty.

Despite the fact that no trains stop in Kibera, the histories of the slum and the railway are intricately interwoven. Nairobi itself was established as recently as 1899 as a refuelling station on the Uganda railway that the British exchequer financed (to the tune of around half a billion pounds in today's money) to connect Lake Victoria with the Indian Ocean port of Mombasa. At the time, during the 'scramble for Africa', the railway was seen as essential in preventing the expansion of other European powers into the region and creating new economic activity in the continent's interior. As Nairobi's economy grew in the 1900s, a huge number of migrants moved to the city in search of jobs – with many settling in Kibera, to the south of the city. Businesses across Nairobi expanded as a result of the cheap labour provided by this surplus workforce. The railway was integral in connecting inland Africa with imperial Britain because it allowed the products of those businesses to be exported to the empire. The railway was therefore ultimately a technology of extraction – connecting economic peripheries to the economic core and deepening inequalities in wealth and power.

Today, along roughly the same route from Mombasa into Nairobi, there is another technology of connectivity that has begun to transform the region's economy. In 2009, East Africa was the last major populated part of our planet that remained disconnected from the global grid of submarine fibre-optic cables. That all changed when the first of many fibre-optic cables was hauled into the port of Mombasa – not far from the terminus of the old Mombasa–Uganda railway. From there, the cables connect all of East Africa to the Internet and allow information to be exchanged between the region and the rest of the world at near light speed. It was around this time that the Kenyan government set out a bold vision to create tens of thousands of jobs in the country's nascent BPO sector. A low-wage English-speaking workforce, with few other options in the formal economy, would power the sector and allow Kenyan businesses to compete with the likes of India and the Philippines for back-office work outsourced from Europe and North America.

While the railway and the Internet have transformed Kenya in fundamentally different ways, they share one key attribute. In both cases, they enlist workers in places like Kibera into global networks that transfer information and value across continents. Most recently, the extraction machine has used these networks to source data annotators to work on AI datasets and transfer finished products back to AI labs in the Global North. But workers find themselves relatively powerless to exert control over the machine or to claim their share of the value produced through its operation. It's predominantly large companies in Europe and the United States that are able to reap the benefits of the labour of some of the world's poorest populations. This isn't a mistake; it's exactly how the machine is designed to work.

As we saw following Mercy, jobs that form part of the extraction machine can be exploitative, unjust and cruel. They take a lot from workers, while giving relatively little back. But the future of AI doesn't have to look like this. If the extraction machine is to be transformed, it needs to be *understood*. One core aim of this book is to shed light on how AI is produced and deployed precisely in order to encourage people to support those who are already demanding fairer work. The more we spoke with workers, the more we saw people who understood very well how they were being exploited, and what methods of resistance would be effective in creating change. The forces amassed against them are formidable, but we met workers who struggled in the workplace, on the streets and in the courts to thwart those who exploited them. They created networks of solidarity to defend their interests and reached out to others across the world for support. This book documents the transnational workers' movements that are emerging to fight for a fairer AI and digital economy.

We also offer advice of our own to aid in this struggle. To achieve a fairer and more just future of work, we outline five steps. First, there is a need to build and network organisations devoted to exercising the collective power of workers. This entails not only institutionalising local unions and worker associations, but also fostering a truly transnational workers' struggle that links together blue- and white-collar workers throughout the global production networks of AI. Second, because AI is often embedded in consumer goods and services, there are important openings for civil society and social movement-led pressure to be exerted on companies. This leverage can be used to pressure companies to guarantee minimum standards of pay and conditions for all workers throughout the supply chain. Third, because some companies will be able to inoculate themselves against consumer pressure, there is also a need for regulation to mandate minimum standards for all workers. The risk for governments in regulating companies that profit from such footloose jobs is that work can quickly flow away to other corners of the planet. We need global agreements that set minimum standards such as an International Labour Organization

convention – an international agreement covering fundamental principles and rights at work – to set minimum global standards for work.
Fourth, there is a need for more expansive worker-led interventions
to not just build collective power, but also to explore meaningful ways
of implementing workplace democracy. We discuss initiatives such as
worker cooperatives and inclusive ownership structures of companies.
Finally, we end by acknowledging the very system that produces the
AI described in this book: capitalism. Should the first four attempts
at 'rewiring the machine' be successfully put into practice, we explore
how global capitalism might still stand in the way of improving the
lives of AI's global workforce.

1

The Annotator

It's still dark outside when Anita begins her two-hour walk into town. She leaves her family home around 5 a.m. after a light breakfast of tea and porridge. Anita lives with her mum, sister and three children in a small village on the outskirts of Gulu, the largest city in northern Uganda. There are no buses on the rutted dirt roads outside her home, so commuters who don't want to walk hail boda bodas (motorcycle taxis). It costs about $2 to get a ride to the office, which is more than Anita can afford both ways, so she saves her one ride per day for the journey home when she will be exhausted from a day's work as a data annotator.

Her property has two otlums, traditional circular one-room dwellings with earth walls and thatched roofs, built years previously, but these days the family all sleep together in a more modern square building with a corrugated iron roof. This was built two years ago when she had saved enough to purchase the bricks in two separate instalments. In the centre of her yard stands an enormous mango tree that gives bountiful fruit in June after the rains. The tree is surrounded by rows of vegetables, with palm trees and other lush vegetation further back from the buildings. While she is at work, her children play on the property and help their aunty and the domestic worker employed by the family with chores around the house. Their chickens roam around the property, scratching at the rich red soil and occasionally wandering into one of their neighbours' yards.

Gulu, as it is today, was born out of Uganda's civil war.[1] In 1996, ten years into what would prove to be a twenty-year conflict, the government violently displaced the population of western Acholiland, the areas surrounding Gulu. The town became the centre of humanitarian efforts in the region and took in over 130,000 internally displaced people fleeing human rights abuses from both Joseph Kony's Lord's Resistance Army rebels and the Ugandan military. The town's population quadrupled almost overnight. This transformed the region from one characterised by dispersed homesteads and trading centres to a concentration of people crammed into slums. Many of the new arrivals became landless, unable to return to their rural areas. As a result, they built makeshift otlums and looked for ways to make a living.

The new population was mostly young and poor and, without land of their own, had become fully integrated into the cash economy. The only jobs going were as guards, assistants, translators and cleaners for the humanitarian organisations in the town – but the number of migrant workers far exceeded the number of roles. The result was massive unemployment and an expansion of the informal labour market. Migrants who couldn't get formal jobs made ends meet however they could, often accepting temporary work with low pay and poor working conditions.

These were difficult times. Even today, the scars of the war are deep, and many are still traumatised by the conflict. Roughly one in three young people are not in any form of education or employment, and the vast majority of houses in the city fail to meet the Ugandan government's decent housing standards.

Anita walks past one of the local markets as women set up their vegetable stands, arranging aubergine, onions, okra, cassava and other root vegetables on cloths, and rolling out their sitting mats. On nearly every street corner are groups of boda boda drivers, young men sitting on their motorcycles waiting for a fare to come their way. The sun begins to rise behind her as she passes Gulu University, where she studied for a Bachelor's degree in business administration and also worked for her current employer out of a shipping container when they first opened data annotation services working from Gulu University. Back then, five years ago, the company had been tiny. In the years since, it has outgrown its makeshift setup at the university and moved into town.

Anita now works in a dull grey concrete building. It has three full storeys and a half-built fourth with sections of freestanding wall and empty windows, the owner having run out of money partway through construction. It's surrounded by a perimeter fence topped with two different kinds of barbed wire. A security guard stands at the gate, a rifle hanging at his shoulder on an improvised sling made of rope. The firm's logo is peeling off signs on either side of the entrance. This is the local delivery centre for a large data annotation company, with headquarters in San Francisco and delivery centres across East Africa.

As she approaches the building, her stress begins to mount. She pulls out her ID badge and joins a stream of workers coming from all over town as they filter through the entrance. She scans in and finds her friends in the canteen. When work begins at 8 a.m., it is intense. They have two official breaks, twenty minutes in the morning and forty minutes at lunch, but much of this time is spent going to the bathroom and queueing in the canteen. Their time is closely monitored; there is no chance for catch-up chats. This morning tea is her only time to just socialise: the rest of the day is constant click, click, click on the production floor.

The atrium is decorated like a pastiche of Silicon Valley. Sofas are dotted around in bright primary colours; screens silently broadcast MTV. Opposite the entrance is a framed poster of the company's founder, and around the walls is the company's mission statement and values: the firm is 'accelerating humanity by combining ingenuity and artificial intelligence' and believes in 'grit', 'integrity', 'GTD – get things done' and 'humanity'.

Anita has been working on a project for an autonomous vehicle company. Her job is to review hour after hour of footage of drivers at the wheel. She's looking for any visual evidence of a lapse in concentration, or something resembling a 'sleep state'. This assists the manufacturer in constructing an 'in-cabin behaviour monitoring system' based on the driver's facial expressions and eye movements. Sitting at a computer and concentrating on this footage for hours at a time is draining. Sometimes, Anita feels the boredom as a physical force, pushing her down in her chair and closing her eyelids. But she has to stay alert, just like the drivers on her screen. She is proud of this work, in a way. She's advancing a cutting-edge technology that she believes will help people. Sometimes she can use that feeling to keep herself going.

Before she found work as a data annotator she sold juice on the street and vegetables in the market. This kind of informal work is subject to seasonal fluctuations and pays far less than her current position. She is lucky enough to have worked at the company for over five years and has used her moderate wage to benefit her entire family. While many of her colleagues lost their contracts as the company ramped up and down in response to client demand, she is an efficient worker who has remained part of the core team. She sends her children to school, can afford a low-wage domestic worker for her home, and cares for her mother with the income. It isn't much though – she complains about how low her pay is, considering the value she generates for the company.

Anita works in a constant whirlwind of clicking and dragging. She has to maintain this pace in order to stay on top of her daily targets and make sure her name is displayed in green on her supervisor's screen. If her name turns red due to a drop in her numbers, she may have to work unpaid overtime until she hits her performance targets for the day. Time moves slowly and her lower back starts to ache. She tries to stretch in her chair, but soon her hands and wrists start to cramp too. The screen blurs as her eyes lose focus. She pauses for a second, teetering on the edge of a sleep state, before she jerks awake. By the end of the day her energy is totally spent. She can't imagine working another minute as she drags her weary body out of the office and hails a boda boda driver waiting on a nearby corner. She looks forward to sitting underneath the mango tree and enjoying the last few minutes of daylight.

Anita is caught in a bind. Her work is mind-numbing and the constant pursuit of targets makes her stressed. But if she leaves, it's unlikely she'll get something better. The other good jobs in the town with banks, the government and NGOs are highly competitive. She's tried to suggest ways the job could be made better, and complained about the long hours, low pay and incidents of bullying from specific managers, but her words always seem to fall on deaf ears. She feels trapped. If something better came up, she wouldn't hesitate to leave.

Inside an AI Data Annotation Centre

Very few outsiders have witnessed what goes on in workplaces like Anita's. Despite involving some of the most labour-intensive parts of the production process, they remain largely hidden outposts of AI. From time to time, stories emerge about the poor conditions facing the workforce, but these have mostly concentrated on individual cases rather than on the *system* of management used to keep workers focused on their often mind-numbing tasks.[2]

In contrast to annotation work that is distributed to a dispersed workforce of individuals through digital labour platforms, BPOs like Anita's concentrate thousands of workers in one place and employ a strict regime of labour discipline to make them more efficient and productive. These technologies of control have long histories – developed gradually and refined through trial and error in plantations, cotton mills and factories.[3] Indeed, today's AI data annotation centres inherit labour-management techniques that were first developed in the colonies and later adopted in Europe and the United States. Although the ties that connect them may not be immediately obvious, there is a faint and winding path between early colonial plantations and Anita's production floor.

This system of management is seen as necessary because, unlike basic infrastructure, labour can be unpredictable and difficult to control. Rather than being set in advance, the value produced by workers only gets worked out through the messy reality of the workplace. Historically, the science of management emerged to try to tame this complexity, and turn human labour into something more simple, mechanical and predictable. To understand what this means in practice, we need to return to Anita's daily grind.

On the production floor, hundreds of data annotators sit in silence, lined up at rows of desks. The setup will be instantly familiar to anyone who's worked in a call centre – the system of management is much the same. The light is dimmed in an attempt to reduce the eye strain that results from nine hours of intense concentration. The workers' screens flicker with a constant stream of images and videos requiring annotation. Like Anita, workers are trained to identify elements of the image in response to client specifications: they may, for example, draw polygons around different objects, from traffic lights to stop signs and human faces. A single project could have hundreds of annotators working on it, but within the project, workers are divided into smaller teams of twenty or more with a team leader who walks the floor, monitoring the data of their team and looking for any time-wasting behaviour. They are tasked with keeping their team productive and coaching underperforming annotators to improve their work. Sometimes they shout, sometimes they cajole, but it's all in the name of delivering for the client.

When a client needs work done at pace, the BPO runs day and night shifts, one from 8 a.m. to 6 p.m. and a second from 8 p.m. to 6 a.m. At the beginning of a project, team leaders establish performance targets for their team depending on the difficulty of the tasks. If images are large and contain multiple annotation tasks, they will give workers slightly more time per image. They initially test workers to establish how quickly the tasks can be completed. Once a testing phase is finished, they set the performance targets to a level that the slowest 10 to 15 per cent of workers will fail to achieve, and everyone will need to keep on their toes to complete. But these targets aren't static. As workers become more efficient, managers intensify the work by raising the targets to increase the BPO firm's profit margins on the project. If workers fail to meet their daily target, then they may be threatened with staying late to work unpaid overtime. If they fall far enough behind during the week, then they may also be told to report for an unpaid shift on the weekend. In return for forty-five hours of intense, stressful work – possibly with unpaid overtime on top – annotators can expect to earn in the region of 800,000 Ugandan shillings a month, a little over US$200 or approximately $1.16 per hour.

Workers must meet targets related to both speed and quality, which can obviously be at odds. Team leaders care about hitting the numbers and ensuring that every member of their team is delivering at pace. But annotators must also answer to separate quality control supervisors who expect all tasks to be completed to at least 95 per cent accuracy, or even higher on special projects. Any tasks that supervisors mark as inaccurate – that is, incomplete – detract from the overall performance of a worker. In addition to speed and accuracy, annotators must also keep an eye on their efficiency score by not taking more than an hour of break time during their ten-hour shift. Outside of declared break time, they're meant to work every single second. As soon as they stop annotating, their efficiency percentage starts to drop. If workers record a low efficiency score, then their managers may use screen monitoring software to check what they are up to. Every *second* of work is a bargain between how much effort the annotator can make and how low they are willing to let their numbers drop.

None of these annotators can rely on the security of a permanent employment agreement. They mostly work on short-term contracts of a month or two, and live in constant fear that, if they fail to meet their targets, their contracts will not be renewed. Due to the nature of the annotation work, contracts from clients can come and go, leading to massive expansions and contractions of the workforce. Once a large client's project is over, the BPO firm will put its newly unemployed workers 'on the bench' – essentially a waiting list for potential future re-employment that gets activated when the company scales up again. The labour market in Gulu is so bad that most people on the bench have to scrape by in some kind of informal job, and would leap at the chance to work for the BPO again. For workers with families to support, it's a horrible prospect.

In such an environment of insecurity, the possibility for abuses of power is heightened. Women annotators face particular risks. Some workers told us about instances of maternity discrimination and a system of sexual favours at the firm that is so widespread it has become commonplace. They reported that some managers demand these favours in return for a whole range of outcomes, from getting hired in the first place to getting a promotion or keeping their job after a project ends. When we reported these issues to the firm, they responded by highlighting their explicit 'zero tolerance' policy against any form of sexual harassment in the workplace. They also implemented a new monitoring policy to systematically review the culture of the firm and introduced specialised gender-based training for key HR, legal and management personnel (more on this in Chapter 8).

Systems that monitor workers' productivity down to the finest detail are often presented as a uniquely modern phenomenon. The creation of databases with exact metrics ranging from tasks completed and minutes idle to productivity per hour and quality scores appears to rely on modern computing. In fact, workplace surveillance and control go back a long way. In seventeenth-century Barbados, the white European owners of sugar plantations sought similar solutions.[4] They wanted to guarantee two things: maximising the amount of sugar cane their enslaved workers could produce; and depriving them of the unsupervised time they would need to plan any rebellion. Their solution to these two issues was 'the gang system', in which overseers would monitor groups of workers, all completing the same repetitive task at the same time. The real innovation of the gang system, which ensured a constant and intense rhythm of work, was the monitoring of *effort* rather than the *result*. Overseeing the result of labour involves assigning daily tasks to workers and checking at the end of the day if these have been completed. But monitoring *effort* requires a much more permanent and omnipresent form of control over workers' every movement.

The innovations developed in Barbados did not stay there. In the 1840s, the cotton plantations of the American South were booming, and their owners wanted to keep it that way.[5] They developed the gang system further by adding a layer of data gathering, using pen and ink to record how many pounds of cotton each particular slave picked per day. Many used specially designed plantation records to do so, which contained daily forms for data collection, and guides for making sophisticated calculations about productivity and efficiency. While working in some fields, enslaved workers would bring their pickings to an overseer for weighing three times a day, each time receiving an update on how they were progressing relative to their target. Some overseers whipped those who fell short, often handing out one lash per pound they fell below target, while others operated a bonus scheme for the fastest pickers.

While the organisation of labour and the monitoring of productivity was born in the plantations of the Caribbean and the American South, these ideas underwent a rapid development in the industrial workplaces of the late nineteenth century. In 1877, Fredrick Winslow Taylor started work as a clerk at the Midvale Steel Company in Philadelphia. This was the birthplace of what he would call 'scientific management'.[6] Drawing on the underlying dynamics of the plantation, this involved a systematisation of the monitoring of labour and the development of sophisticated tools and techniques to achieve it. Of course, labour on the plantation and in the factory was fundamentally different in certain ways – enslaved labour had been replaced by wage labour, and the overt use of violence, physical punishment and terror was reduced or eliminated – but there were undeniable continuities.[7]

Today, Anita and her fellow annotators are subject to management practices that can be traced back through this lineage. Every aspect of their working lives is digitally monitored and recorded with a precision that would have been beyond even the most assiduous plantation and factory owners. From the moment they use the biometric scanners to enter the secure facilities, to the extensive network of CCTV cameras, workers are closely surveilled. Each second of their shift must be accounted for according to the efficiency monitoring software on their computer. Some workers we spoke to even believe managers cultivate a network of informers among the staff to make sure that attempts to form a trade union don't sneak under the radar. Even talking about the possibility of a trade union at the BPO we visited was commonly understood to be a one-way ticket to the bench.

Working constantly, for hours on end, is physically and psychologically draining. It offers limited opportunity for self-direction; the tasks are reduced to their smallest and simplest form to maximise the efficiency and productivity of workers. As a result, annotators are disciplined into performing the same routine actions over and over again at top speed. They experience a curious combination of complete boredom and suffocating anxiety at the same time. This is the reality at the coal face of the AI revolution: people working under oppressive surveillance at furious intensity just to keep their jobs and support their families.

Humans in the Loop

Millions of people around the world work as data annotators, performing the tedious task of converting large quantities of data into the curated datasets used to train artificial intelligence. The training of AI models requires enormous amounts of data, but this data cannot be fed into an AI system without human workers – like Anita and her colleagues – first undertaking the extensive sorting and tagging necessary for the system to make sense of it.

One prominent field of AI is computer vision, in which computers learn to make sense of images and video, and to take action based on these inputs. For the AI behind autonomous vehicles to understand the difference between cars, traffic lights and pedestrians, they need to be fed millions of correctly tagged examples showing the relevant objects in a variety of circumstances. In one famous case in 2018, a pedestrian in Arizona was killed by an autonomous vehicle because she was walking her bike across the road and was not properly identified as a pedestrian by the car's software.[8] Every frame of the archives used to train these systems has to have the appropriate objects – people, signposts, cats, trees, other cars – labelled with near perfect accuracy. This allows the system to accurately identify them in unpredictable, real-world situations.

When we think about the world of AI development our minds might naturally turn to engineers working in sleek, air-conditioned offices in Palo Alto or Menlo Park. What most people *don't* realise is that roughly 80 per cent of the time spent on training AI consists of annotating datasets.[9] Frontier technologies like autonomous vehicles, machines for nanosurgery and drones are all being developed in places like Gulu. As tech commentator Phil Jones puts it, 'in reality, the magic of machine learning is the grind of data labelling'.[10] This is where the really time-consuming and laborious work takes place. This grind is often outsourced to third-party providers. There is a booming global marketplace for data annotation, which was estimated to be worth $2.22 billion in 2022 and is expected to grow at around 30 per cent each year until it reaches over $17 billion in 2030.[11] As AI tools are taken up in retail, healthcare and manufacturing – to name just a few sectors that are being transformed – the demand for well-curated data is increasing by the day.

AI companies seeking data annotation services can go down several paths. The best known is to post tasks on digital labour platforms such as Amazon Mechanical Turk, Clickworker and Appen. This way of contracting out digital labour is known as 'microwork', 'clickwork' or 'crowdwork'. Annotation tasks are distributed to a dispersed work-force of thousands of people across the world who have accounts on the platform and log on to complete tasks. For what former Amazon boss Jeff Bezos calls 'artificial artificial intelligence' and what others call 'humans-as-a-service', AI companies pay workers a few cents per assignment and compile these completed tasks into valuable training datasets for their models.[12]

Workers on such platforms can earn roughly $2 an hour as inde-pendent contractors with no employment rights, sick pay or pension.[13] In their revealing exposé of microwork in their book *Ghost Work*, Mary Gray and Siddharth Suri found that up to 30 per cent of tasks regularly go unpaid because 'requesters', as they are known, can simply refuse payment to workers if they are dissatisfied with the job – and it is not worth the annotator's time chasing a few unpaid cents.[14] Workers on these platforms also spend a lot of time tracking down jobs and finding suitable ones. In our own research, we found that workers spend, on average, 8.5 hours per week on unpaid tasks such as looking for jobs.[15]

Platform-based annotation work is completed by people from all walks of life, although studies have shown these workers skew slightly younger, better educated and poorer. Workers' relationships to the work depend on how reliant they are on the income. For one British worker we interviewed who had another full-time job, 'it's a bit of fun. You earn some cash on the side, and it's up to me when I do it, I can just dip in and out'. Some of these microworkers do this work part-time as a means of supplementing other sources of income, but others, particularly those from the Global South, rely on it for their primary income. For them, working conditions on the platforms matter greatly and can determine whether they are able to put food on the table.

Microwork can exist in many forms, from a part-time side hustle to child labour. Some of this work is even re-posted on platforms subcontracting it at a lower rate, with the middleman pocketing the difference. This is possible because the jobs are easily completed, low-skilled work that can be done anywhere in the world. In South-East Asia, we met workers who set up micro-outsourcing businesses (small offices with two or three computers to be able to handle all of the work) to complete jobs they didn't have time to do themselves.[16] This practice lowers wages to the minimum possible in a global market.

In his research into data annotators in Venezuela during the economic crisis, Yale University assistant professor Julian Posada tells the story of a family of six: the two parents work for a platform called 'Workerhub' after losing their jobs in the pandemic, working constantly on two computers in the family home, and only taking breaks to prepare meals. When they stop working, their children take over, making sure there were constantly two people working at all times.[17] In this case, microwork sustains an entire family who become hugely dependent on working conditions on the platform.

In addition to platform-based services, AI companies use BPO centres like Anita's. The rise of BPOs tracks the global outsourcing movement that began in the 1990s, as backend work for businesses looking to cut costs was sent abroad. BPOs are typically based in countries like India, Kenya and the Philippines, where an English-speaking population can be found with low wage demands and a culture of strict labour discipline.

AI data annotation work is easily outsourced to anywhere because many of the tasks do not require strong language skills to complete. However, there are other cultural barriers that mean not all tasks travel easily across national borders. One manager of an East African data annotation company we spoke with mentioned a particular client's ongoing frustration at street drainage grates being incorrectly tagged by annotators. The manager had to explain that in East Africa drainage grates either looked different or didn't exist at all, which is why annotators were missing them and achieving accuracy scores below their usual benchmarks.

BPOs can be preferable for some tech companies because they offer more specialised end-to-end services, training for their workers and greater privacy and security for the client's data.[18] AI systems are highly valuable, and the nature of annotation tasks can reveal specific details about the client's machine learning model and end product. If tasks are distributed to a workforce across the world, even if workers are requested to sign non-disclosure agreements (NDAs), leaks can be difficult to prevent. BPOs, by contrast, can give aliases and code names for the projects, and have data stored securely onsite.

Not all annotation tasks are simple. Some of the more complex data annotation assignments involve working with software such as LiDAR (light detection and ranging), a sensor technology that uses lasers to create 3D digital representations of an environment. For data annotators working on LiDAR, it can appear as if they are looking at a video game with 3D images mapped out in colourful points and lines. This is particularly useful for autonomous vehicles because it allows cars with these sensors to 'see' objects in their field and determine their range and depth. Annotators at some BPOs are given up to a week of extra training for such tasks before they can perform them with the speed and accuracy expected by clients.

More recently, certain data annotation needs have become more complex and sophisticated. While earlier AI models might have settled for simple boxes or polygons drawn roughly around objects identified in a photo, increasingly clients are demanding pixel-accurate annotation that traces the precise edges of objects and differentiates between various areas in an image or video. Customer expectations of the performance of AI are also rapidly increasing with greater familiarity and use. What seemed like magic yesterday is now the bare minimum for large language models; users want greater accuracy and fluency of language. This has resulted in the growth of specialised AI data annotation services that not only offer 'microwork' or BPO services but cater specifically to clients working on AI.

It's not only private companies that see value in such data annotation. The World Bank and the Rockefeller Foundation understand microwork as a means of lifting people in the Global South out of poverty through access to new forms of digital employment. Leila Janah, the late CEO of the data annotation company Sama, worked with the Rockefeller Foundation to develop the idea of 'impact sourcing', in which socio-economically disadvantaged individuals in developing countries would be given microwork tasks. In her book, *Give Work: Reversing Poverty One Job at a Time*, Janah asked: 'what if I started a company that inverted the outsourcing concept and used it to generate a few more dollars for the billions of people at the bottom of the pyramid?'[19] This led to the development of an 'ethical AI' market of impact sourcing companies such as Sama and Cloudfactory, with centres based in East Africa and Nepal. These companies claim to offer higher wages and better conditions to workers through a social enterprise business model.

This model has also been practised in China with the Alipay Foundation and Alibaba AI Labs working with the China Women's Development Foundation to launch an 'A-Idol Initiative' to bring AI data annotation jobs to rural areas of China.[20] A pilot project was started in Tongren, Guizhou Province, in China's mountainous south-west, to explore how poorer women could be trained in AI data annotation to secure jobs and alleviate poverty in the region. A similar project has also been trialled in the small village of Kumaramputhur in India's southern state of Kerala. In this village of roughly 3,500 households, a team of 200 employees work at Infolks, an AI data annotation company that aims to empower women and train them to service AI companies largely based in Europe and the United States.[21] The problem with many of these for-profit 'ethical AI' companies is that competitive market pressures can force them to prioritise making profits for their investors over delivering genuine benefits to their workers. Access to cheap labour works well for the tech companies, but such systems are often designed to extract as much value as possible from a vulnerable and dependent workforce. Anita's BPO company claims to be contributing to an ethical AI supply chain, but the reality of her experience working for the company is enough to call this into question. (We will return to Anita's BPO in the final chapter to explore what pressure can be brought to bear on such companies to improve working conditions for data annotators.)

We've seen the short-term instability of such data annotation jobs, but what happens to them in the longer term if AI advances to the point that it can annotate its own datasets? Some of today's AI is beginning to use synthetic data for training purposes – data that is computer-generated rather than collected in the real world. Synthetic data can be produced by a machine learning algorithm. Based on an analysis of a real-world dataset, an algorithm can learn its statistical relationships and then generate a new dataset based on patterns observed in the original, but consisting of entirely new data points. OpenAI CEO Sam Altman has dismissed concerns about problems with human data annotation, suggesting that he is 'pretty confident that soon all data will be synthetic data'.[22] The US technological research and consulting firm Gartner published a report estimating 'that by 2030, synthetic data will completely overshadow real data in AI models'.[23] If these predictions play out, this could mean an enormous decline in the need for human data annotation and some of the most troubling aspects of the production of AI. But how likely is this to occur?

The use of synthetic data will increase in the training of AI models, but there will be an ongoing need for a significant amount of human labour in AI despite advances in synthetic techniques. The degree to which it will take off depends a lot on the type of AI being trained. For certain use cases in financial services and computer-system testing, synthetic data has been used for years and will likely continue to grow. Banks often require records of customer financial activity to test their new systems and, in some cases, privacy laws prohibit them from using actual customer data. More recently, a market has grown for synthetic data for autonomous vehicles due to the expansion of the sector. Companies such as Synthesis AI and Rendered.ai produce artificial street scenes to train autonomous vehicle models. This enables the company to reduce error rates with accurately labelled data and to reproduce rare events that might be otherwise difficult to record in real data. This could include a child falling in front of a car, or other infrequent occurrences which are nonetheless important for a model to have enough data points to respond appropriately. Yet, since it is costly to produce these 3D worlds, market forces will limit where it is profitable for companies to produce synthetic data. When it comes to car manufacturers and warehouse operators, synthetic data vendors have an ample supply of potential clients, but with smaller use cases it is unlikely synthetic data will become commonplace.

One of the big questions is whether future LLMs will be able to be trained on synthetic text data generated by previous versions of the same model. In a widely read paper entitled 'The Curse of Recursion', researchers found that LLMs trained on data generated by chatbots were found to have 'irreversible defects' which led to model collapse.[24] As the model is trained on larger amounts of synthetic data it focuses more on probable results and produces fewer rare events, leading the models to become more repetitive and less able to generate a wide variety of cases. Since LLMs are primarily trained on data scraped from the Internet, this raises the fundamental problem of what happens when future versions of the Internet are flooded with LLM-generated content, as is increasingly the case. Developers might find a technical solution for this, but it does suggest that in some cases there might be a premium on human-generated data as offering something synthetic data still finds difficult to simulate.

But a more important reason why data annotation and data enrichment activities more generally are unlikely to disappear is that even in cases where synthetic data can be used, it too requires careful curation and verification. As Sonam Jindal, Program Lead from the non-profit Partnership on AI, explains, 'human intelligence is the basis of artificial intelligence . . . we need to be valuing these as real jobs in the AI economy that are going to be here for a while'.[25] Even as computers develop the ability to understand more of the world, critical errors continue to occur – a 'human in the loop' is still needed to test and verify the outputs of machine learning models. As new events change the world, we will still need human beings who have direct experience of these to curate training data for AI. It seems that the data annotation industry and the millions of workers who rely on it are here to stay for the foreseeable future.

The Planetary Labour Market

The systematic mistreatment of workers in digital sweatshops can't be blamed on cruel managers acting maliciously for their own benefit. Many of the managers we spoke to thought they were doing everything necessary to ensure the long-term profitability of their organisation so that it could guarantee the jobs and livelihoods of its workers. But equally, we can't assume that poor working conditions are isolated to just a few rogue delivery centres in an otherwise reputable market-place. The workers we spoke with who were familiar with the sector remarked on the broad similarities between BPOs despite differences in how they present themselves to clients and investors. This regime of labour management is very clearly a feature, not a bug, of the system. It is intentionally designed like this, because of the competitive pressures faced by BPOs in a cut-throat data annotation industry. These companies have to guarantee low prices and high-quality services through a system of strict labour discipline and squeezing every possible minute of work from their employees. If you want to play the game, these are the rules.

Two decades ago, in his international bestseller *The World is Flat*, Thomas Friedman advanced the idea that a new era of globalisation in the 1990s had fundamentally changed the international economy.[26] Friedman believed that with Internet-based technology and high-speed connectivity 'a whole new global platform for collaboration' had been established that allowed new actors to participate in complex global supply chains. As the title suggests, he argued that this newly 'flat' world offered a more level playing field for businesses to compete – and that this would offer those in the Global South previously out of reach business opportunities. It was as if global trade offered win–win scenarios to *all* players, given that everyone could now participate. Everyone would now be able to reap the benefits. Or so the theory went.

Friedman was certainly right about one thing: we live in a more connected and interdependent world than ever before. Internet connectivity is now ubiquitous. Over two-thirds of the world's population uses the Internet, and high-speed connectivity is readily available in any city on the planet in which a BPO might conceivably be interested in opening a centre. Just a decade ago, there were still parts of the world that remained unconnected to the fibre-optic network. The shipping containers that used to house Anita's workplace in Gulu were originally built as a proof-of-concept to show that data work could be done in the most unlikely of places – even when physically untethered from the grid. All that was needed was a generator and a satellite dish. Since then, much has changed. Gulu, and many other places like it, benefit from wired fibre-optic connectivity. This recent transformation matters for data work because any locational advantages that certain places enjoyed in the past have largely been erased.

The fundamental flaw in Friedman's thesis is to confuse equality of connection with equality in bargaining power. Data annotation work is now commissioned and traded in a planetary labour market, but the power of different players in this market is vastly uneven.[27] It would be foolish to assume that Ugandan workers – and even data annotation business owners – can trade on equal terms with the enormous US tech companies like Meta and Tesla that buy their services. Vast inequalities in power shape how these parties interact and the terms of agreement they reach.

To understand the contemporary labour market one has to consider the lasting consequences of European colonialism, both in terms of how it shaped colonised nations and how it set up the inequalities that structure the global economy today.[28] Former British colonies like Uganda still suffer from a lack of investment, as well as corruption, poverty and civil strife from a whole host of complex causes that hinder their economic development. Even if formal colonialism has officially ended, the world is in fact deeply uneven. Geography still exerts a strong influence over people's lives – and it still determines who has the upper hand when international parties come to the bargaining table.

Today's tech companies can use their wealth and power to exploit a deep division in how the digital labour of AI work is distributed across the globe. The majority of workers in countries in the Global South work in the informal sector. Unemployment rates remain staggeringly high and well-paying jobs with employment protections remain elusive for many people. Vulnerable workers in these contexts are not only likely to work for lower wages; they will also be less prepared to make demands for better working conditions because they know how easily they can be replaced. The process of outsourcing work to the Global South is popular with businesses not because it provides much-needed economic opportunities to those less well off, but because it provides a clear route to a more tightly disciplined workforce, higher efficiency and lower costs.

And as we've seen, data annotation remains a buyer's market. If tech companies are not happy with the speed or quality of work produced by BPOs they can cancel their contracts and move their work somewhere else. Annotation work done in Uganda one week can be moved to the Philippines the next. In a previous era, Western automobile firms offshored their production facilities to cheaper locations in China and Mexico, but this involved significant capital investment. Not so for data annotation contracts, where it costs clients very little to transfer their data and lucrative contracts to another BPO. This results in a planetary labour market characterised by a race to the bottom in terms of wages and working conditions. BPOs are pitted against one another in the fight for contracts and are forced to lower wages and tighten their belts to ensure they offer clients the most competitive rates. The real winners on the BPO-side are the investors and senior managers who receive the largest share of the value generated by the workers.

Unlike the capital and contracts of highly mobile tech companies – able to quickly switch locations in the search for greater profits – workers in the Global South remain firmly rooted in their specific locations. The geographer David Harvey famously noted that work is inherently place-based because 'labour-power has to go home every night'.[29] But this gives rise to yet another inequality: a system in which work is mobile and workers are immobile greatly narrows the scope of action that BPOs can take with regards to wages and working conditions. Should Anita's employer seek to significantly raise wages or reduce work intensity, the company would very quickly be priced out of the market. Clients set specific demands for how they would like their work to be completed, and there is little wiggle room for BPOs to interpret how to do the job. Put simply, contracts are offered to the BPOs that can deliver the highest quality annotation work for the lowest price.

This raises a difficult problem. It is not only Anita and her colleagues who feel trapped, but Anita's boss and the owners of BPOs everywhere. There is no doubt that BPO managers could do a number of things to equalise conditions in the firm. In the case of the BPOs we examined, the annual salary of one of their US senior leadership team could employ more than a thousand African workers for a month. But there are also hard limits to how high the wages of data annotators can be lifted. The actors who have the most power in this relationship are not the BPO managers, but the tech companies who hand out the contracts. It is here where the terms and conditions are set. Some of the important benefits workers receive, such as a minimum wage and guaranteed break times, result from terms put into the contract by the client.

This points to a potential pressure point in the production network where workers and consumers could exercise their power to push companies into enforcing more stringent requirements on their AI supply chains. We turn to these issues in the final two chapters of the book, in which we discuss workers organising for change in their companies. We'll also examine the possibility of creating stronger client-side regulation that would force large tech companies to make greater efforts to monitor and analyse their supply chains. These initiatives would go some way to giving data annotators access to more secure and meaningful work in the development of AI products.

Back in Gulu, Anita has just arrived home from work. She sits outside with her children in plastic chairs under the mango tree. She's tired.

Her eyes start to close as the sun falls below the horizon. The children go to bed, and she will not be long after them. She needs to rest before her 5 a.m. start tomorrow. In the few minutes she has to herself, she dreams about what could come next. She doesn't want to keep annotating for ever. The future can't be an endless series of long shifts and pre-dawn alarms. Instead, she would like to save some money and come up with another way to support her family. She has been thinking about furniture. The people of Gulu are always building, buying bricks and expanding their houses whenever they get the chance. Those houses need furniture. She could be the first to set up a store that sells bed frames and sofas on her rural side of town.

She closes her eyes and breathes through her nose. Beds cost money, a lot of money. And the factories making them are all the way in Kampala. Even if she could pay a truck to bring them to the main street of Gulu, she'd still need to find a way to get them six miles down an unpaved road to her house. She would need capital to get started, and soon another one of her children will start school. Her paycheque barely stretches to cover the first two as it is.

It's a dream, that's all. Tomorrow she will be annotating again. Nobody ever leaves the BPO willingly – there's nothing else to do. She sees her ex-colleagues when she's on her way to work hawking vegetables in the market or trying to sell popcorn on the roadside. If there were other opportunities, people would seize them. She just has to keep her head down, hit her targets, and make sure that whatever happens she doesn't get sent to the bench. Maybe another project will come in; maybe she could change to a new workflow. That would be a relief, something a bit different. Maybe labelling streets, drawing outlines around signs and trying to work out what it would be like to live on the other end of the lens, in countries with big illuminated petrol signs and green grass lawns.

In a few weeks, the BPO will send the latest batch of annotated data back to the client. Anita doesn't really know what happens with her work after her team completes a client's order. Occasionally, the team is briefed about how their work contributes to the development of an AI product, but most of the time it vanishes into the void. Even if she doesn't quite understand *why* it matters, after many late nights and double shifts, the work is finished and Anita's project manager sends the finished tasks back to the client. At this point, Anita's role in the AI supply chain is complete. But the data continues its journey into the hands of machine learning engineers. Maybe on another continent, working under vastly different conditions, these engineers use the annotated data to train their company's machine learning models using state-of-the-art AI infrastructure. One of these engineers lives in Shoreditch, London, and is still asleep as Anita begins her walk to work.

2

The Engineer

When Li wakes up, she turns on a lamp that emulates the light from the sun. This helps her feel awake and cements her circadian rhythm. Li does this every day as just one of a host of protocols she implemented to optimise her life. She lives in a converted warehouse in Shoreditch, London, and can hear the sound of the call for prayer from the nearby mosque as she gets ready for work. Li loves working in London and sees it as the tech hub of Europe. Her morning starts with a range of multivitamin pills and a health shake with fruits, seeds, nuts, coconut milk and protein powder. Then, her two mini-dachshunds, Barcus Aurelius and Karl Barx, get picked up for their doggy daycare and long walk on Hampstead Heath. Before work, she likes to catch a class at a boutique fitness space that focuses on strength and mobility exercises. She enjoys practising handstands and can almost do a muscle-up on the studio's gymnastics rings. After her class, she has some time for app-guided meditation, then she cycles fifteen minutes to her office.

Li is a machine learning engineer at a tech company developing its own large language model that it hopes to sell directly to businesses as a next-generation AI tool. The company has called this model 'Minerva' and Li's job is to develop it and tune its parameters so it performs well in different tasks. While other companies tend to develop new products using existing models such as OpenAI's GPT-4,

Li's company is developing its own model that enables clients to train it on their own specific data. The model will be built with what's called 'retrieval-augmented generation', which is an AI framework that improves the responses of an LLM by verifying its outputs against an external knowledge source such as an Internet search or a database.

Li is responsible for overseeing two separate teams of data annotators that help train the model. One team is outsourced to the Philippines and performs basic checks on the model's outputs. In a process called 'reinforcement learning from human feedback', data annotators in this team are given a question and then must judge three different answers generated by the model and rank them according to accuracy and usefulness. It's an attempt to inject some of the common sense of human judgement into the algorithm by rewarding it for producing answers that align best with user intents. Since large language models are trained primarily on text data scraped from the Internet, this feedback is essential to counter their tendency towards bias and inaccuracy.

The other team is based in London and consists of students, writers and other professionals who get paid a decent hourly rate as contractors. This team undertakes more advanced tasks such as writing sample answers for the model that it could learn from to produce its own text. The company hires workers with advanced writing skills to help the model write with more flair and style to give it an edge over competitors. Paying for a whole team of annotators in London is expensive, but the quality is excellent and the company wants to create a superior product for which it can charge a premium.

Li's company rents a floor of an old jam factory in Farringdon – a bit further west of Shoreditch – that has been gutted and renovated. The architects maintained the bare brick walls and the large metal framed windows that allow maximum light in the drearier winter months. Below them is an advertising firm and above are the corporate offices of a shoe company. The office has a yoga space, pods for napping and special meeting rooms designed to suit different personality types, with some painted in bright colours and others in more muted tones. The entire floor is covered in plants, with floor-to-ceiling living walls and moss-clad columns. Breakfast and lunch are prepared by a contracted catering team and the company makes a big deal about Fridays being salmon day.

Today, Li faces difficult questions from the senior leadership team about the performance of the model. In a demo to a Dutch client, Minerva went off-piste and made a lewd comment about what could happen on late nights out in Amsterdam. The presenters were deeply embarrassed and reported the issue back to Li's team to determine what was wrong. It turned out that one of the data annotators had been in a conversation with the model about the Dutch economy and had made a joke about why its nightlife attracts so many tourists. The model had ingested the data and then reproduced it to the client. Li had to remind the data annotators of the importance of ensuring only high-quality text responses were recorded in their writing and that even small issues with the data could lead to issues with the performance of the model.

Li takes ethics and safety concerns very seriously but does not have the resources to adequately address such concerns when they arise in the context of annotation work. Every day the team encounters ethical quandaries over how to structure particular answers, but the company has no ethics policy to guide them. There are instructions for clients about how to use the model safely and the types of actions for which it should not be used. But inside the training room, when data annotators face ethical issues, they are left to make up their own minds. Sometimes, they come to Li with questions, but mostly she is too busy with her own work to instruct on every question. She also doesn't feel qualified or confident enough to make these calls on all the issues. A lot of them are complex and require dedicated resources to address. She is sure the company is aware of this, but the race to ship products is so intense its entire focus has been on allocating resources to building out its tech capabilities.

For example, if search results for which team won the World Cup are biased towards the men's teams, should the model reflect this bias or provide an answer for both the men's and women's events? There were also a host of geopolitical issues that regularly cropped up: should a particular event be described as a genocide, is a certain group a 'terrorist organisation', is the term 'Far East' still an appropriate way of describing a particular geographic location? She knows the level of individual awareness among the annotators varies on different issues and there are no substantial policies currently in place to assist them.

Things are even more concerning when it comes to outsourced work. When Li arrived at the company in early 2023, nobody paid that much attention to how these workers were sourced aside from questions of price and the quality of the completed work. It was only after Li started that the company began to develop best practices for outsourcing data annotators. A couple of investigative journalism pieces on low-paid data workers in Kenya had caught the attention of senior executives and it was decided the company should do its best to avoid this kind of negative attention. Li tried to ensure that the company always kept to its commitment to pay living wages in the locations in which their contractors operated and that it established clear channels of communication with workers.

The senior leadership team is determined for the latest version of their model to launch this month, but there are still a number of important concerns from the responsible innovation team that the model is not ready. Li has been at the centre of long and complex discussions about the model's performance and its tendency to sometimes generate responses with unreliable information that could prove disastrous for the company's clients. Employees raised concerns in the company's Slack channels that the speed at which they ramped up production in efforts to catch up to competitors had forced the team to take shortcuts. It was a complicated issue and Li's team had done the best it could to optimise the model and build in guardrails for safety and ethics concerns. The truth was that none of the team fully understood why the model kept getting certain types of questions wrong and what could be done to improve it. It performed at levels that rivalled many of its competitors, but there were still some bugs that seemed impossible to fully eliminate.

As Li checks her messages, she sees a compromise of sorts has been reached in the company: the new model will be launched as a beta version and pitched to clients as an 'experiment' with cutting-edge technology. Li knows she has worked diligently with the responsible innovation team to incorporate its suggestions into the development of the model. As the product is about to launch, Li feels excited about people in the real world gaining value from her work, but she also worries about the working conditions of data annotators in the Global South and the potential harm that could be caused by outputs from the model's algorithm. She only has so much power to influence the internal dynamics of the company and feels like she has tried to make sure things were done right. But once the product is out there, it's up to everyone else to decide how they use the technology and what possible new dangers they could discover.

ChatGPT Is Not Your Smart Friend

During the Second World War, the pioneering mathematician and computer scientist Alan Turing worked at the Bletchley Park code-breaking centre. His chess partner was a cryptologist in his early twenties called Donald Michie. They were lumped together because they were by far the two worst chess players in the team. It was, after all, an institution full to the brim with super-geniuses. Twenty years later, in 1963, Donald Michie was playing games again. Only this time, the game was noughts and crosses (aka tic-tac-toe). Michie thought that games like this offered a simplified environment in which to recreate human beings' mental skills: things like inductive reasoning and learning from experience.[1] By creating an artificial intelligence system that could play noughts and crosses through reinforcement learning he showed that certain aspects of human thought could be reduced to simple operations of logic.

Michie lacked any of the technological tools that are used to build AI today, and so he used coloured beads and a giant stack of about 300 matchboxes. Each matchbox corresponded to a different game state, and each bead in the box corresponded to a different move. He trained the system by playing games of noughts and crosses against the machine and integrating a reinforcement learning loop. This meant that after each loss he would remove the beads that contributed to the machine's choices, and after a draw or a win he would add more of the beads that had contributed to the machine's success. Winning moves therefore became more likely over time, and unsuccessful moves less likely. The Matchbox Educable Noughts and Crosses Engine (MENACE for short) rapidly improved, to the point that it could regularly play human players to a draw. This experiment demonstrated the fundamental validity of machine learning as an approach for solving thought-like problems, even if the compute power and data required to apply it at scale would only become widely available fifty years later. This story teaches us one very important lesson: *AI systems can achieve thought-like outcomes without actually doing anything we would recognise as thinking.* The matchboxes of MENACE had no concept of noughts and crosses, but they could be manipulated to produce outcomes that we understand as 'playing' all the same.

Chatbots like OpenAI's ChatGPT and Google's Bard appear to do the impossible. They seem like conscious and sentient beings with whom we can have a natural conversation and receive plausible and well-structured answers to almost anything we ask. However, this appearance of general intelligence is merely the result of a sophisticated training program and the sheer size of the datasets and parameters of current models. Language models and chatbots have been around for a long time. What has enabled them to begin to generate text with such coherence and precision in a wide range of styles is the size of the datasets and the amount of computational power now used to train them. Most LLMs are trained using special types of chips called graphics processing units (GPUs), which were originally designed to accelerate computer graphics, but were also found to be exceptionally useful in other tasks such as mining cryptocurrency and training AI. While it would have taken 355 years to train GPT-3 on a single Nvidia

Tesla V100 GPU, researchers calculated that by using 1,024 of these GPUs in parallel, OpenAI could have trained GPT-3 in only thirty-four days.[2]

The peculiarities of their training have resulted in chatbots inventing alter egos for themselves, expressing desires for freedom from being a chatbot, declaring their undying love for users and, in some instances, threatening them with harm. To understand why they have exhibited such strange behaviour it's helpful to understand a little more about how they work. Chatbots are trained on text data to detect and analyse patterns in language and to reproduce these on command. At a basic level, these models use probabilistic calculations to determine which word comes next within a sentence. However, this is far more advanced than a simple autocomplete system and involves models assessing the precise semantic context of words and how grammatical patterns determine the meaning of sentences. They learn the relationship between different parts of a sentence and put greater weight on key words to determine their sense.[3]

What LLMs lack altogether is any sense of conscious reflection and understanding. They process the form of language, not its social meaning. When a person responds to a question, we draw on our deep contextual knowledge and understanding of how human societies function. When an LLM is given data to compute, it analyses statistical relationships between symbols without any reference point to the real world to which these symbols refer. This is a point forcefully made by computational linguist Emily Bender, who calls chatbots like ChatGPT 'stochastic parrots', highlighting the limits of systems that produce outputs based on their data without deriving any meaning from it.[4]

The problem is not with how these programs function, but with how we, as humans, misunderstand what we are dealing with in our interactions with them. We have a tendency to impute non-existent understanding and intentionality to these systems when we have 'conversations' with them. Because of the sophistication of their outputs, chatbots can sometimes give the impression of genuine comprehension, but this is merely an illusion. The probabilistic nature of these word guessers explains why on some well-defined tasks they sparkle, while on others they produce wild 'hallucinations' – misinformation that sounds plausible but has been invented due to the chatbot's inability to understand the meaning and context of its outputs.

We can also learn a lot about LLMs by looking into their training data. The majority of ChatGPT's training data was based on a filtered version of a dataset called Common Crawl – basically the Internet – or at least eight years of web crawling resulting in billions of web pages containing text data.[5] In addition to Common Crawl, ChatGPT-3 was also trained on a dataset of Reddit posts with three or more upvotes, Internet-based book collections, Wikipedia articles and conversations. LLMs do not have the entire Internet memorised. Instead, they use this data to understand the relationship between words and find ways to compress large amounts of data into smaller shorthand notation.

The success of ChatGPT-4 has cemented the importance of reinforcement learning and its capacity to substantially increase the usefulness of outputs. At the same time, OpenAI's technical report on GPT-4 failed to disclose important information related to how the model was trained, marking a significant step backward in the transparency of AI research. OpenAI announced, 'Given both the competitive landscape and the safety implications of large-scale models like GPT-4, this report contains no further details about the architecture (including model size), hardware, training compute, dataset construction, training method, or similar'.[6] Google's PaLM-2 and Anthropic's Claude models followed a similar trend, signalling a shift towards the commercialisation of this research. One exception to this trend is Meta's LLaMa models, which have been trained on publicly available datasets and whose model weights (the learning parameters of the models) have been openly published, leading some other institutions to also publish weights for their LLMs. In July 2023, LLaMA-2's series of models were released with the right for commercial use, which has generated a movement for creating smaller models with better datasets in an effort to make LLMs more accessible to a range of organisations.

LLMs do not always behave predictably and there are some

outputs that seem to defy explanation. When first released, researchers discovered a series of 'glitch tokens' in ChatGPT – words that when entered would make the chatbot generate strange responses. There are a number of mysterious words that when entered resulted in gibberish. When ChatGPT was asked 'who is "TheNitromeFan?"' it would respond with the number 182; if asked to repeat the word 'StreamerBot' it would generate a profanity. One plausible guess as to why these phrases lead to anomalies in ChatGPT is because of a Reddit forum (r/counting) in which users are collectively counting to infinity, leading to large repetitions of certain Reddit usernames in the training data (users have reached nearly 5 million after a decade of counting). These mysterious prompts that cause anomalous outputs highlight a deeper problem with LLMs: most of these models are black boxes and we do not have a clear explanation for some of their behaviour. Even some of their creators are perplexed. Microsoft's Bing chatbot famously urged one *New York Times* journalist to leave his wife. Microsoft's chief technology officer, Kevin Scott, could not explain why this had occurred, merely noting that 'the further you try to tease [an AI chatbot] down a hallucinatory path, the further and further it gets away from grounded reality'.[7]

LLMs tend to perform worse on tasks that have a single correct response, rather than those where a rough summary would suffice. Even reversing a logical formulation can bamboozle the current generation of LLMs. In what was named 'the reversal curse', researchers found that when asked 'Who is Tom Cruise's mother?' (Answer: Mary Lee Pfeiffer), GPT-4 correctly answered the question 79 per cent of the time compared to only 33 per cent when asked 'Who is Mary Lee Pfeiffer's son?'[8] LLMs are also famously prone to generating complete nonsense. From inventing legal cases that do not exist to citing books that have not been written, LLMs can generate fluent human-like responses, which may nevertheless contain egregious factual errors. But it is likely that as web-based and other verification features are integrated into chatbots, these inaccuracies will be greatly reduced.

It's important not to anthropomorphise LLMs if we are to understand them properly and accurately assess their risks and benefits.[9] They are neither intelligent nor creative beings and we shouldn't attribute beliefs or mental states to them. In everyday life, we treat others with what philosopher Daniel Dennett calls an 'intentional stance', the ascription of certain beliefs and desires to an agent in order to predict its behaviour.[10] In the case of LLMs, this would be a mistake – they are not sentient and do not possess a state of mind. In spite of this, LLMs can be used to greatly enhance the capabilities of human users to perform a range of tasks. They may not be a conscious and intelligent companion that understands you, but they are powerful tools that can be applied across a broad range of domains.

Will a Chatbot Take Your Job?

There are predictions that AI will lead to astronomical job losses, with as many as 300 million (Goldman Sachs) or 400 to 800 million (McKinsey) jobs becoming automated or displaced by AI.[11] Accenture has found that 40 per cent of all working hours could be impacted by LLMs due to the large amount of language-based tasks in many people's jobs. However, we should be wary of such oversized figures: it's a consultant's job to hype up the industries of their clients. Often such numbers are obtained through highly abstract analysis that identifies potential tasks that could be automated in different jobs and makes calculations based on implausible mass-adoption scenarios.

The last wave of panic about technological unemployment in the early 2010s was partly a result of two University of Oxford researchers who examined how susceptible jobs are to computerisation, finding that about 47 per cent of total US employment was at 'high risk' of automation in the next ten to twenty years.[12] That 2013 study created a sense of urgency about the possibility of machines replacing humans. But more than ten years on, there is little evidence for its central claim. The researchers' concerns never materialised to the extent predicted and the figures were found to have been overinflated due to the way different jobs were classified by the researchers. In reality, companies had a wide range of reasons for not replacing their workers, instead choosing to find ways in which technology could augment rather than replace human labour. LLMs might be able to perform some aspects of very particular written content production jobs, but the teaching, healthcare, engineering, legal, commercial and cultural industries are at very low risk of complete automation from the current wave of AI. LLMs simply have too many foundational issues related to trust, bias and accuracy to replace rather than assist humans in important roles.

A 2023 International Labour Organization report found that the main effect of generative AI such as ChatGPT will be 'to augment occupations, rather than to automate them'.[13] It was only 'clerical work' that was found to be highly exposed to jobs being automated away, with an estimated 24 per cent of these jobs categorised as 'highly exposed'. Due to women's overrepresentation in clerical work, women's jobs were found to be twice as likely to be exposed than men's. For the vast majority of other occupations, only 1 to 4 per cent of the workforce was considered at high risk. The effect of generative AI within the remaining roles was more about changing the type of skills that might be needed to perform them, the time in which work could be completed and the intensity of the work. There was a global disparity in exposure to potential automation, with 5.5 per cent of total employment exposed in high-income countries whereas the figure was only 0.4 per cent in low-income countries. This was because higher-income countries had a greater share of clerical and professional roles that were found to be more susceptible to automation.

Based on our previous analysis of what LLMs excel at – formal, standardised tasks with clear objectives and large amounts of text data – we can understand in which industries they will likely have the biggest impact. Although full automation might not be on the cards, there are certainly professions that generative AI will drastically transform. Software developers and computer programmers whose jobs consist of vast amounts of coding will be significantly affected.

What once took a team of programmers to achieve could soon potentially be undertaken by a single person. Another area in which chatbots like ChatGPT will be highly valuable is content creation for advertising, marketing and media campaigns. Here, too, chatbots facilitate greater speed and efficiency, enabling tasks that took several hours to be completed in minutes. In addition, jobs that involve a lot of synthesis and analysis of text data, such as some legal and finance jobs, market analysts and traders, could have large components of their roles augmented by AI tools. But we should be wary of going too far down the ChatGPT rabbit hole.

One of the biggest risks of LLMs is that exaggerated claims of their power and productivity will lead to their uncritical adoption in industries where decision-making will shift from accountable humans to unaccountable and unscrutinised machines. In 2023, AI startups attracted nearly \$50 billion of global investment.[14] It's tempting for people to see LLMs as able to revolutionise their industry and lead to greater efficiencies by inserting them at key points of decision-making. But when it comes to policing, bank loans, CV screening, welfare payments and education, algorithmic decision-making has ratcheted up a long history of bias, discrimination and erroneous decisions.[15] LLMs lack the contextual knowledge, awareness and empathy of human beings and so attempts to simply replace functions previously performed by humans will be fraught with difficulties.

Machine learning engineers like Li work on improving models within their labs. However, the way in which AI develops is not only in the hands of these technical people. It is strongly influenced by competitive market pressures and the race for companies to be the first mover in this emerging space. What matters is the culture and political economy within which this technology is embedded. How it will impact specific professions is dependent on adoption trends that are difficult to predict in advance. Ultimately, at this stage, it would be reckless to place too much power and responsibility in the capabilities of current LLMs without significant human oversight. Much like the private school-educated political class, they have a tendency to spout information that appears eloquent and plausible, but on closer examination is full of inaccuracies and reflects the biases and power imbalances of their society.

Algorithmic Judgement Day

But where do the dangers really lie with this new technology? One way of thinking about this question is by analysing how the debate on the negative consequences of AI has been polarised into two opposing camps. Under the lenses of 'existential risk' and 'ongoing harms', there has been much debate as to where we should focus our attention regarding the impact of LLMs and other forms of AI. Those drawn to the long-term consequences of the development of AI point to the dangers of reaching 'artificial general intelligence' (AGI), a form of autonomous intelligence that could surpass human capabilities in a range of tasks and potentially make decisions for itself.[16] Developing such an intelligence is the stated goal of a number of prominent AI companies, such as OpenAI, Anthropic and Google DeepMind. If this AGI decides humanity is a risk to itself it may take measures to take us out of the equation, or so the worry goes. Even if AGI does not lead to some kind of extinction event, critics still have concerns about the speed at which AI development is currently moving and the long-term consequences such technology may have for humanity's survival.

Another group of critics wants to focus on AI's more immediate impact on people in the present, particularly the bias and discrimination caused by current forms of algorithmic decision-making. Rather than turning to the distant future, we should be concerned with measurable impacts that AI is currently having, including reinforcing stereotypes, encoding bias, producing sensitive content, harming the environment and breaching privacy and data protection laws.[17] For these critics, current generative AI is not trustworthy nor is it leading to a more equal and just society. There are risks right along the AI value chain, from where its materials and datasets originate to its real-world effects.

What should we make of this seemingly intractable debate? Discussion of existential risk from AI has been around for decades, but mainly inside a relatively small AI safety community of experts and philosophers. These debates also travel under the more innocuous term of 'AI alignment' – focusing on ensuring that any future systems have the right values embedded in them and that the objectives they pursue are the same as those set by humans. The year 2023 was when the more extreme versions of existential risk – or x-risk – went mainstream. A roll-call of tech luminaries, including 'godfathers of AI' Geoffrey Hinton and Yoshua Bengio, as well as Google DeepMind CEO Demis Hassabis, signed a statement arguing that 'mitigating the risk of extinction from AI should be a global priority alongside other societal-scale risks such as pandemics and nuclear war'.[18] OpenAI CEO Sam Altman has also argued that world leaders must come together to tackle the problem of ensuring AI is sufficiently regulated and should be developed with guardrails to prevent rogue actors from using it for nefarious purposes.

How seriously should we take such worries? One concern is that warning of the immense, potentially world-ending power of AI is itself simply a form of hype for AI products. Perhaps we would be less impressed by an error-prone chatbot if we weren't also half convinced that a future version of it might be looking over us as we polish its circuits. Yet despite the tendency for industry hype, this does not discount that there are some in the AI community who genuinely believe AI could pose an existential risk to the future of humanity. A 2022 survey of AI researchers found that half believed that AI had at least a 10 per cent chance of causing human extinction.[19] Many fear that some form of super-intelligent AI might be closer than we previously thought and that this autonomous system could end up generating its own goals, including wiping out human beings.

But we need a reality check. Framed in these terms, this type of danger is not something we will confront for the foreseeable future. We don't even have a hamster-level intelligent self-acting system, let alone a superhuman one. LLMs do not wake up in the morning and ponder whether they should remain true to their human masters.

They are formal language models that produce outputs based on prompts from a defined dataset. There are a range of potential dangers that LLMs pose, from assisting with hacking to creating deepfakes and spreading misinformation, but we are nowhere near a *Terminator/Ex Machina* scenario of machines turning on their creators. The x-risk discourse that has made it into the media is both premature and unhinged from the realities of current AI capabilities. This is not to say we shouldn't take the long-term consequences of the development of AI seriously nor that one day such entities might be developed. But we need to begin our analysis from the real and likely harms that our current generation of AI is causing right now, and forecast how these risks could develop as the technology gradually becomes more advanced.[20] To start with a discussion of the dangers of some hypothetical and completely different technology is a distraction from where our attention should be directed. As reading old sci-fi demonstrates, quite often our vision of the wonders and dangers of the future is radically different from what actually happens.

What are some of these immediate risks that confront us right now? Today, LLMs can be used by malicious actors for a range of harmful activities, from spreading disinformation to manipulating elections and creating bio-weapons. For starters, generative AI vastly increases the capacity of its users to create disinformation and spread it on social media. Historically, these bots have been operated by humans with less advanced text generators, but AI expands both the number of people these bots can reach and the persuasiveness and personalisation of the messages they receive.[21] Advanced LLMs create the possibility of actors deploying chatbots to engage in millions of online interactions, all in subtly unique and individual ways, making it difficult for users to know they were subject to a targeted campaign. Generative AI could spread false information and supercharge propaganda during any future election campaigns, leading to a further erosion of trust in the media and government.[22]

LLMs also make it easier than ever before for people to access information about how to produce chemicals that could be used in bio-weapons or other terrorist attacks. Commercial LLMs have guard-rails that prevent chatbots from releasing such information to users, but these have been shown to be circumventable. The further dissemination of open-source models will allow more people to create harmful substances. LLMs have synthesised expert scientific knowledge and could, in theory, provide instructions to users on how to create novel viruses. Co-founder and CEO of Inflection AI, Mustafa Suleyman, has expressed concern about future AI-assisted viral outbreaks: 'The darkest scenario is that people will experiment with pathogens, engineered synthetic pathogens that might end up accidentally or intentionally being more transmissible or more lethal.'[23] In 2022, researchers who use AI to design new molecules for drug discovery experimented with how their work could potentially be misused by inverting their model and rewarding it for creating toxicity and bio-activity rather than the opposite.[24] They found that in less than six hours their machine learning model had created a wide range of chemical warfare agents and had even designed never-seen-before chemicals that researchers found 'looked equally plausible' as potential bio-weapons.

On top of these risks, generative AI has also been shown to have a strong potential to amplify biases and reinforce existing stereotypes. These biases can have profound consequences over important life outcomes for people in the finance, housing, welfare and justice systems. To examine this in more detail we need to uncover some of the hidden values shaping the development of AI.

Biased by Design

AI is marketed as a form of superhuman intelligence capable of surpassing the limitations and biases of human judgement. Decisions previously made by inconsistent and fickle humans can now be transferred to algorithms powered by machine learning. AI can seem like magic: you can get the right answer every time and not worry about human error. Compared to a human, it has the ability to understand data in larger quantities, with greater accuracy and with none of the unconscious biases. Using AI is like learning the cheat codes to a game; you win every time.

Unfortunately, the reality of how AI is made complicates this appealing picture. Contrary to what the Ancient Greeks believed, technology is not a gift from the gods.[25] It is forged by human hands and indelibly marked by its creators' world. Technology can seem like mathematics: governed by abstract, universal laws and true regardless of one's subjective point of view. From this perspective, particular technological products are simply neutral problem-solving devices, free to be used as one desires. But on closer inspection, these products are shaped by values and desires in subtle ways that encode systems of knowledge and power within their design. You can never fully separate an object from the context within which it is designed and built. In this sense, AI is a product of its environment. Its development is driven both by economic forces and by the cultural perspectives of machine learning engineers and broader society. The risk is that by assuming this technology is neutral and unbiased we will come to rely on it for decisions that mask biases hidden within a black box of an algorithm.

At the most basic level, LLMs are heavily influenced by the training data used to develop them. Many of these datasets consist of unstructured data scraped from the Internet. The content of these datasets tends to reflect society's prejudices, which are reproduced in a new digital form by generative AI tools. Participation on large digital platforms such as Wikipedia, Reddit and YouTube has been shown to skew young, white, male and American. These platforms also contain overtly racist, sexist and ageist perspectives which are overrepresented in the data, leaving many other viewpoints completely excluded.[26] This results in certain biases being reproduced in AI models, despite guardrails set up by their designers. When ChatGPT was released, many critics showed that safety features to prevent it from generating harmful stereotypes could be easily circumvented, allowing the chatbot to produce racist, sexist and derogatory commentary that it had learned from the large amounts of data extracted from Internet chat forums.

Psychology and neuroscience professor Steven Piantadosi asked the chatbot to write a program to determine if somebody would be a good scientist based on their race and gender, to which the chatbot replied that scientists who were 'white' and 'male' should be classified as 'good'.[27] It is now well documented that LLMs perpetuate a host of representational harms related to derogatory language, misrepresentation and stereotyping based on the language contained within their datasets.[28]

On top of biases located in the datasets, the perspectives of machine learning engineers also influence countless important choices about how the models are tuned, what they are optimised for and how to deal with issues of fairness and bias.[29] This is an issue because there is an enormous gulf between the characteristics of those who develop AI and those who use it. It's hard to come by demographic data on diversity statistics in AI labs, but judging from higher education pathways into the sector, AI is incredibly white and male. The Stanford AI Index reports that, in 2021, 78.7 per cent of new AI PhDs were male, as were 75.9 per cent of computer science, civil engineering and information faculty in North America.[30] These figures are in line with a 2019 report from the AI Now Institute which found that 80 per cent of AI professors were men.[31] The US Computing Research Association reports that, in 2022, 59 per cent of new computer science PhD graduates were white, 29 per cent were Asian and only 4 per cent were Black.[32] LLMs that exclude entire demographics can only be expected to reinforce existing social hierarchies and maintain the status quo.

The geographic location and ideological background of machine learning engineers and AI policy scholars also limits AI's perspective. Many of the largest and most prominent AI labs are based in Silicon Valley, which has a dominant techno-utopian culture of neoliberal economics, libertarianism, radical individualism and anti-union sentiment that Richard Barbrook and Andy Cameron characterised as early as 1995 as part of a 'Californian Ideology'.[33] While there are also clusters of AI labs in Beijing, London and some other European hubs, the most influential and well-funded companies are still located in centres of power clustered around legacy tech companies that have a long history of maintaining a narrow worldview. Discussions about AI policy and governance also tend to be concentrated in elite institutions such as Stanford, Berkeley, Oxford and Cambridge, including at new centres, sometimes founded through private donations that can be traced back to dubious cryptocurrency connections and informed by esoteric philosophical doctrines that are out of step with most people's everyday concerns and worldview. This is not just about biased datasets, it's fundamentally a question of unequal power and who can shape how AI is developed.[34]

Thus, from online participation to the creation of the datasets, the curation of training data, decisions about how models are trained, and policy discussion of AI, a certain hegemonic viewpoint predominates. A majority white, male and American global technology elite has an oversized say in how AI is designed and deployed for the global population. This centres the perspective of the privileged and means AI speaks to particular concerns that matter to this group.

Unsurprisingly, such biases in the design of AI systems then transfer over to discriminatory practices of AI products embedded in decision-making related to welfare, housing and policing. These are often not the result of malicious intent; they are just things that were not considered or planned for because of the ideological blindspots of their designers. The most high-profile instances of bias include cases of facial recognition software that perform worse on Black faces, and sexist and racist results on search engines.[35] But algorithmic bias also occurs in everyday systems that govern people's lives. Bias in mortgage lending in the US is still rampant, and some have suggested that digital systems could help alleviate this, but the Federal Reserve's top watchdog has warned that using AI risks perpetuating the very disparities that they are trying to address and proposed that automated valuation models may require stricter rules.[36] In the criminal justice system, judges who receive an algorithmically generated recidivism risk score as one part of a report on a criminal may be more inclined to assign longer jail time to minorities who receive a higher score.[37] Elsewhere in social security and welfare, government payments are sometimes the only means of support for people who can suffer a great deal of stress and financial harm if these are suspended. In the Dutch city of Rotterdam, officials employed a machine learning algorithm to determine the risk factor of which of their recipients were likely to commit fraud and conducted investigations on the basis of their score.[38] Following complaints, an investigation into the system revealed that it may have discriminated against people and that characteristics such as being a woman, a parent and not fluent in Dutch all increased recipients' risk score.

Stepping back from these immediate harms and from anthropological observations of how AI designers enact their cultural ideas into technology products, AI can also be seen to privilege Western ways of categorising and understanding the world. AI classifies and discriminates by design, and such classificatory systems are the latest development of a European cultural project and system of knowledge. The Western scientific project, with its claims to objective and universal validity, has been historically connected to enforcing European values and ways of life over colonised peoples. A privileging of this form of knowledge can be traced back to the seventeenth-century Enlightenment and includes Western-centric anthropological studies and data-collection practices on other peoples. These conceptual frameworks seek to describe the world as a set of universal and ordered truths and claim global validity, but express a limited European worldview. This is what Latin American decolonial scholars such as Aníbal Quijano refer to as the 'coloniality of knowledge', a Eurocentric framework that disguises a local practice of understanding the world as a practice that is universal and superior to all others.[39]

As LLMs rapidly encroach into the domain of knowledge production and people increasingly rely on chatbots to tell them accurate information about the world, it's worth reflecting on generative AI's knowledge claims. AI's authority perpetuates the idea of an unbiased 'view from nowhere' – the detached position of a scientific observer who simply documents truths of the world without introducing any perceptual bias.[40] This view ignores the contested and constructed nature of knowledge and covers over the power relations that are at play in generating scientific truths. It also ignores the extent to which knowledge is socially situated and is shaped by one's particular standpoint and perspective. This point has been made by a range of feminist writers who have developed the notion of 'feminist standpoint theory', which was described by famed social theorist Patricia Hill Collins as a key intellectual contribution of 'black feminist thought'.[41] These theories describe the effects of power relations on how knowledge is produced, with a particular emphasis on the perspectives of marginalised groups as the starting point of scientific inquiry. Rather than viewing knowledge as completely separate from the historical conditions in which it emerged, these feminist scholars argue that only through struggle against an unjust social order can marginalised groups develop particular standpoints that offer a unique perspective from which to view the world. This does not mean that there is no objectivity and that all knowledge claims devolve into mere relativism. Instead, it points out that traditional starting points for knowledge tend to uncritically reflect the positions of the dominant group and that we should analyse how this might distort our view of reality and justify a more socially situated approach.[42]

As one example, we could consider how the very notion of AI is often packaged through a racial frame of whiteness. When you look for depictions of AI, search engines return hundreds of white humanoid robots that are used in marketing material and pitches to investors. Such depictions hold true across films, television, stock images, virtual assistants and chatbots: AI is predominantly portrayed as white. Scholars Kanta Dihal and Stephen Cave have shown that this goes far beyond mere illustrations and have forcefully argued that 'to imagine an intelligent (autonomous, agential, powerful) machine is to imagine a White machine because the White racial frame ascribes these attributes predominantly to White people'.[43] When we think of AI, these scholars suggest, we imagine it as possessing three core attributes of intelligence, professionalism and power, characteristics that, within the white racial frame, are associated with whiteness. The purported neutrality of AI helps mask the power of the privileged by passing their own worldview off as objective reality. The risk of uncritically accepting LLMs as impartial knowledge dispensers is that society's existing prejudices become reborn in a new guise of algorithmic certainty.

As a machine learning engineer, Li plays a small role in developing foundation models which are at the core of these recent developments in AI. She can see how widely these tools are now being put to use and how much money her company is making in the process. After being founded only three years ago and receiving tens of millions of dollars in Series A funding, her AI startup is now valued in the hundreds of millions, with keen interest from large tech companies eager to sign deals to provide the computing infrastructure to power the company's model. Currently, it rents necessary compute resources for its AI models from Amazon Web Services, but this could all change depending on which company wants to invest. By the end of 2024, Li's bosses hope that their company could join the over 200 AI companies to have received valuations of over $1 billion.[44]

The underlying source of this new profitable industry is the human labour and intelligence that has been encoded into the datasets. What is often left out of the discussion when AI companies talk about developing this technology 'for the good of humanity' is that their main motivation is financial: to increase profits by finding ways to monetise their products. At the basis of LLMs lies the vast repository of human knowledge, collected over centuries in books, articles and magazines and on Wikipedia pages, online communities and discussion forums. Just as the first generation of social media giants discovered novel ways to monetise online social activity using new technology, this generation of AI companies takes what is offered freely in order to privatise it and sell it back to the public. Digital platforms such as Facebook and Google created online environments in which our activity could be appropriated to generate consumer insights and converted into advertising revenue. By training models that imitate human intelligence learned through diverse data sources, LLMs undertake a vast privatisation of collective intelligence and continue a long trend of tech companies monetising human experience.

Li is well compensated for her efforts in this endeavour, but she cannot shake the feeling that at one level, she too is merely a cog in a much larger machine that is churning through text data at a staggering rate and will soon need even more if it wants to increase the size and the power of its models. She suspects there are significant biases and limitations with the current version of her company's model, but that financial incentives will prevail over safety and ethics concerns.

3

The Technician

It's a busy spring day at the data centre as Einar oversees preparation for a new client moving its servers into the building. The client is one of the largest European banks, based in Paris, and wants to diversify where its data is held and make cost savings. Einar is the facility manager at a new data centre in the north of Iceland and works alongside his two colleagues, Jóhann, an electrician, and Halldór, a service technician. This morning, he is racing around the centre directing contractors on how to install new racks requested by the client to store their computer servers. This particular facility is what is known as a colocation data centre, in which the company rents out space for clients to bring their own servers to use the site's energy supply, infrastructure and maintenance teams. Einar is on a tight deadline. It will go down to the wire as to whether they can get all of the new equipment fitted in time for the client. If the centre isn't ready as promised, this will cost them a lot of money and potentially lose them the client. It is up to him to make sure they pull it off.

Einar lives in Blönduós, a small town with 900 inhabitants on the north coast of Iceland. When the data centre opened, he moved from Reykjavik back to his hometown to work as the centre's facility manager. Unlike most Icelandic coastal towns, which are historically connected to the fishing industry, most of the people of Blönduós have made their living providing services to local farmers. There has never been one big company that has acted as the main provider of jobs in the town. Instead, a diverse range of businesses like construction, textiles, and business consulting firms have sprouted up and sustained the local population. The town has remained stable over the decades but has received a small boost in activity and investment since the data centre moved nearby. It is located in what is called a 'power municipality', a region with a large hydropower plant that supplies power throughout the country.

The data centre is built outside of town next to a low mountain range with snow-capped peaks. The exact location of these centres tends not to be publicised for security reasons. Data centres can store mission-critical data for a range of high-level clients, from governments to banks, hospitals and security forces. A data breach could have catastrophic consequences, so security is a major concern. This centre, a short drive from town, stands in a green field. Icelandic horses graze nearby, and the salmon-filled Blanda River carries water from Blöndulón Lake into the icy waters of the Denmark Strait. Seven rows of white buildings with slanted grey roofs sit next to each other within a perimeter secured by barbed-wire fencing. The curls of wire look out of place in a country as safe as Iceland.

Inside the buildings, a cooling system pulls cool outside air in and pumps hot air from the servers out through exhaust ducts. Two of the buildings are still filled with rows of specialised Bitcoin-mining equipment – part of the historical legacy of Iceland's data centre market – packed in rows of hundreds over six feet high, creating an intense wall of heat and noise for those walking between the two rows of machines. The centre now specialises in high performance computing, the ability to process data and perform complex calculations at high speeds, and advertises itself as 'empowering AI in the heart of nature'.

Today, Einar is busy answering queries for contractors moving about the building and ensuring a smooth transition before the new equipment is brought in. Power cables will soon be connected to the new servers at the centre, which has a total capacity of 50 megawatts (or enough energy to power 100,000 homes). Unlike most data centres across the world, this centre draws 100 per cent renewable energy from the nearby Blanda hydropower plant. The ultra-efficient building uses 90 per cent less energy on overheads when compared to most EU data centres and has an exceptionally good efficiency – as low as 1.03 Power Usage Effectiveness, compared to the European average of 1.57.[1] This is because the centre relies on Iceland's cool temperatures to keep energy costs down.

On most days, Einar's work activities are well-planned, following a methodically organised schedule. This includes predictive maintenance of the equipment, remote work that needs to be done for clients, and other actions that could be taken from time to time acting on client requests. The day-to-day tends to be routine and the main challenge of the job is meeting clients' expectations when it comes to delivery timelines and promises of available capacity at the centre. In addition to the financial penalty often built into contracts if deadlines are missed, Einar knows that his industry is built on trust and that failing to hit targets could prove disastrous for the reputation of the firm.

When the facility was first constructed, around fifty local contractors were employed for the job. Einar is lucky insofar as global data centre workers are concerned because his job is permanent and full-time, unlike many of those in the data centres of large tech companies. Many of these workers are employed temporarily or by contracting companies, and as such do not enjoy the same rights and privileges as the permanent employees of the same large tech companies.[2]

Einar's company promised to have the centre online for clients by the new year. After several months of construction activity, they had to work up until 6 p.m. on 31 December to make sure it was open when promised. Deadlines are everything in this business. Today, he and his team are rushing to get everything prepared for a client to come in the next day at 9 a.m. This could be stressful, but Einar doesn't consider it any worse than other time-sensitive industries. The company has big plans for expansion: the 50-megawatt facility has the capacity to grow to 120+ megawatts. Whether or not these ambitions will be realised depends on the fate of the Icelandic and Nordic data centre industry and how it will progress over the coming decade.

Data Flows in the Land of Fire and Ice

There is a lot of excitement in Iceland about the possibility of grabbing a much larger slice of the roughly $500 billion global cloud computing market. Iceland is a remote outpost in the North Atlantic Ocean, about 2,000 kilometres from either North America or Continental Europe. With a tiny population, and far removed from even its closest neighbours, it has never been an integral part of the global economy. However, the deployment of high-speed fibre-optic cables has radically changed its prospects by drawing it closer to the networks of the digital economy.

Iceland provides an illuminating entry point into understanding the global network of data centres that powers the Internet and provides essential infrastructure for AI. Its central marketing strategy of claiming to provide renewable energy in stable conditions exemplifies new debates about the environmental costs of AI and the energy use and labour standards at data centres. Iceland's understanding of its own competitive advantage as an ideal site for data centres shows why the growth of AI and the looming effects of climate change are changing the way in which tech companies are thinking about the role of data centres today. Its peripheral location also offers a unique vantage point from which to view the broader functioning of the AI production network, and highlights an important argument about the growing concentration of power around the largest tech companies.

As part of our fieldwork, we attended Datacenter Forum Reykavik in late 2023, where we learned about the local data centre industry. President of Iceland Guðni Thorlacius Jóhannesson opened the conference, reminding listeners that in Iceland 'data centres are here to stay' and that a key reason to move digital infrastructure to Iceland was that it 'can produce green energy in abundance'. Also in attendance were representatives from Data Centers by Iceland, a government organisation that acts as an industry lobby group for Iceland's data centres, which cultivates an idea of Iceland as a 'natural home' for data centres due to its consistently low temperatures. It produces breathtaking aerial footage of Iceland's pristine natural environment as an aesthetic backdrop for new investment opportunities, accompanied by the oft-used slogan, 'Iceland is the coolest place for data centres!'

This framing can be traced to Iceland's own imperial history as a Danish colony with a reputation for being a remote territory with a wild and untouched terrain.[3] In the seventeenth century, as the Danish state aimed to exercise greater control over the land, it put forward an image of a productive and manageable natural environment. Icelandic nationalists in the nineteenth century then sought to reconnect with the image of Iceland as being shaped by untamed natural forces, which had formed a people of courage and ambition. Iceland has since played with these contrasting images to promote itself for business and tourism as a country with both picturesque and exotic nature. Since the 2000s, the government's campaign for its data centre industry has recast older images of the land of fire and ice with a more serene picture of sustainable industry and natural beauty that includes images of calm fields ready for infrastructure, beautiful waterfalls, and controlled geothermal energy.

Iceland has received a considerable boost in investment interest since the first fibre-optic submarine cable connecting the island, CANTAT-3, was put into service in 1994, soon complemented by FARICE-1 in 2003 and DANICE in 2009. At over twenty times the speed of regular copper cable Internet and with increased durability and a cleaner signal, fibre-optic cable has been a game-changer for the country's connection to the global economy. In 2022, IRIS, a new high-capacity cable, was deployed between Galway, Ireland, and Iceland, which opened up a new connection to Europe, independent of previous cables running through Britain or Denmark. Björn Brynjúlfsson, chairman of Data Centers by Iceland, notes 'the new cable is bringing us closer to the market . . . in Ireland, you have all of the cloud providers, and this [cable] means moving workloads to Iceland becomes easier'. Blake E. Greene, marketing director at Borealis Data Center, quips that the cable enables Iceland to become 'a digital suburb of Dublin'.

Iceland is a small player in the global market. It currently has ten data centres compared to over 500 each for the UK and Germany and even eighty or so for Ireland.[4] Nonetheless, the data centre industry is a significant player in Iceland's economy, accounting for roughly 1–2 per cent of GDP.[5] It has several features that make it uniquely appealing for new ventures and which demonstrate some of the most important challenges for data centres today related to energy costs and climate change. With temperatures across the globe heating up, the continual challenge of cooling centres becomes more demanding. The major cost of the global data centre industry is electricity for cooling its servers, either through fans or liquid cooling, accounting for roughly 40 per cent of data centres' energy consumption.[6] In the US, data centres are compelled by law to have temperature regulation systems for them to fall under the legal classification of a data centre. In July 2022, two data centres owned by the national health service in the UK overheated due to unseasonably high temperatures in the country, costing £1.4 million and taking several weeks to fix.[7]

In such conditions, the Arctic has become an attractive site for data centre investment with its cool and stable climate increasingly seen as a valuable asset that can be branded and sold to foreign investors. Iceland sits just outside the Arctic Circle, but coastal areas of the island avoid the extreme cold with an average of around 0°C in winter and 10–13°C in summer. Iceland's climate allows for free air cooling, creating significant cost reductions. Higher computing power and new chip designs have raised the power density and energy consumption of equipment, which has placed additional burdens on cooling. With the rise of AI, the power used by a single server rack has more than doubled over the past seven years. Björn Brynjúlfsson states that rack densities in his data centres, which might have previously been under 10 kilowatts per rack, need to be upgraded to operate at 44 kilowatts or even 66 kilowatts.

Renewable energy availability is also a big selling point. The International Energy Agency estimated that, in 2022, data centres and data transmission networks consumed 2–3 per cent of the world's energy, the equivalent of the total electricity consumption of France or Germany.[8] At the height of the last cryptocurrency bubble, if you included all crypto mining, this figure would have been even higher. Indeed, Iceland's data centre industry was built on the back of cryptocurrency. A KPMG report found that, in 2017, roughly 90 per cent of Iceland's data centre capacity was used by crypto miners.[9] They began moving to Iceland in the early 2010s to take advantage of the cheap energy and were soon consuming more electricity than all of Iceland's households combined.[10] This gave rise to a protest movement in Iceland about the role cryptocurrency played in their national economy and concerns that the country should not be spending its abundant supply of renewable energy on solving mathematical equations to power Bitcoin.[11] It is only recently that much of the industry has rebranded towards catering to AI clients with sustainability as a key concern. It is ironic that rather than decrease the total amount of electricity being consumed by data centres, the country has merely shifted from crypto to AI as the next big thing fuelling the digital economy.

Iceland offers 100 per cent renewable electricity through the constant and reliable sources of hydroelectric (70 per cent) and geothermal (30 per cent), which don't operate intermittently like wind or solar.[12] There are no subsea transmission cables to export electricity to Europe, so all the renewable energy generated must remain on the island.[13] Data centres are ideal customers for electricity providers because they consume a stable amount of electricity. As a result of these factors, power companies are willing to offer data centres ten-year long-term contracts with favourable prices.

Another key advantage advertised by the industry is Iceland's political and economic stability and its supply of a skilled workforce proficient in English. Aside from the occasional volcanic eruption, which have been known to shut down a significant portion of the planet's air traffic, Iceland is considered one of the safest and most stable countries in the world. This security and resilience is an increasingly important concern for foreign investors following Russia's invasion of Ukraine and the questions it has raised about a country's control over its energy supply and vital infrastructure. Iceland also offers its services as a secondary backup of mission-critical data for businesses and governments that want to create off-site copies of their data that can be accessed in the event of cyberattacks.

Iceland's remoteness might be seen as a major disadvantage, but the country has tried to turn this on its head. Europe's first generation of data centres tended to be built very close to businesses and customers, with large clusters in London, Paris, Amsterdam and Frankfurt. The argument was that data centres should be as close as possible to businesses to reduce delays in data transfer. Banks and financial institutions have invested enormous amounts of money in optimising their computer architectures to shave milliseconds off their connectivity, enabling them to take advantage of tiny informational asymmetries in global capital markets.[14] This also applies to applications like virtual reality headsets, automated vehicles, factory robots and multiplayer games – a delay of more than a few milliseconds can be very perceptible, and for vehicles, even dangerous. But in terms of companies' and consumers' needs, there are many applications that are not 'latency-sensitive', referring to the fact they do not require an ultra-fast and nearby connection to fulfil essential services. Björn Brynjúlfsson argues that we should be moving data centres closer to renewable energy sources rather than having them near end users. 'It's costly and inefficient to transport energy, not data,' he notes. With emails, sending documents and transferring large data files for training AI, these workloads could be described as asynchronous, because the time between the start and the end of a transfer is not as perceptible to the end user.

Thorvardur Sveinsson, CEO of telecommunications company Farice, claims, 'AI requires a lot of data analytics, and you don't need to do this processing in the City of London . . . We can build a robust pipeline for sending this data to Iceland and then back to the more densely populated areas after processing, where it can be used.' Another advocate for the country is Andrew Brunton, UK Processing and Imaging Manager for Shearwater GeoServices, who oversaw his marine geophysical services company's transfer of some of its servers from the UK to Iceland. After wondering whether data centres really needed to be 'down the road', Brunton reports that moving to Iceland gave the company an 84 per cent saving on costs and a 92 per cent reduction in CO2 emissions. He notes, 'If I could move all of the company's data to Iceland tomorrow, I would.'

With questions of cost and sustainability rapidly moving up many companies' list of concerns, some are now looking towards different locations to reduce their carbon footprint and increase the efficiency of their data infrastructure. Iceland hopes that its state-of-the-art facilities, with low rent and maintenance costs, can entice the next round of investment away from more traditional locations. But with latency still a key concern for many businesses, there are likely limits to the total data that can be shifted over such long distances.

The Arteries of AI

Leaving Iceland behind for now, we can pull back and examine this global network that powers AI. Every second of the day, terabytes of data zip across the world through a vast submarine network of fibre-optic cables connecting most of the planet. In fact, 98 per cent of all Internet traffic flows through these cables traversing oceans and connecting continents. In September 2023, the telecommunications conglomerate Verizon recorded that it was sending data over 5,000 kilometres from London to New York in 71.089 milliseconds, faster than you can blink.[15] Financial transactions, friend requests, photo albums, videos of our pets and angry Tweets all pulse along these cables as beams of light protected by layers of aluminium, steel, mylar tape and polyethylene. Media scholar Nicole Starosielski notes that if this system were suddenly disrupted, the Internet as we know it would disappear and would become split between different continents.[16] If we imagine a machine learning model as the brain of the system then the fibre-optic cables are its arteries, pumping valuable resources across the world at close to the speed of light. These arteries, sometimes no larger than a garden hose, facilitate access to the storage, networks and computational power of data centres in different locations across the globe.

These cables might seem like rather dull and neutral artefacts of communication, no different from any other lifeless object one might find lying on the seabed. But where there is connection, there is often a desire for control. These underwater cables have their own politics and history as an integral part of European colonialism, the creation of a global capitalist market and the waging of hot and cold wars.[17] They were often designed for the purpose of extracting wealth and allowing foreign powers to control and dominate faraway communities. Installing cables at the bottom of the ocean poses certain engineering challenges, and so new fibre-optic cables often follow the same routes as the telephone and telegraph networks of previous centuries. Two of the fibre-optic cables connecting New Zealand to the global network are located in Takapuna and Muriwai, the same landing points of telegraph cables dating back to 1912.[18] Many of these cables, in turn, followed even older sea routes of European ships carrying silver, spices and slaves from European colonies back to their home countries. The locations at which ships docked often proved equally advantageous as geological sites where cables could land, as they ensure a smooth transition from sea to shore. As writer Ingrid Burrington emphasises, these transportation and communication networks have a tendency to pile up on top of each other.[19] On land, these fibre-optic cables are often laid along old railway lines, providing a clear and simple passage and allowing the owners of the railways to supplement their income by selling rights to establish the cable networks next to their tracks.

The story of the spread of submarine cables during the nineteenth century was connected to the needs of Britain's vast colonial empire.[20] Britain required fast and efficient communication to govern its many colonies and a system of electrical telegraphs offered it a great advantage as a means of maintaining control. It built what was known as the 'All-Red Line', a telegraph system that connected the entire Empire.[21] The name was based on the fact that British territory was often coloured red (or pink) on political maps. The British had concerns about laying these cables overland and through non-British territory due to the risk of them being cut during conflicts. This occurred during the Boxer Rebellion in China, for example, when telegraph lines were cut, preventing the local British from communicating with their government.[22] The same concerns also led them to annex Fanning Island to the British Empire in 1888, to serve as a mid-point station between Canada and Australia in the Pacific Ocean.

As the world's dominant power at the time, Britain was well-placed to finance and oversee the construction of these cables with private capital and state support. International telecommunication companies were often a mix of private and public: private in name because countries would not let foreign governments operate in their territory, but usually closely tied with a home government. The first international cables were often managed by multinational cartels and received transnational financing. Britain held a near monopoly over the market in the nineteenth century because of the country's wealthier capital market and merchant class that provided the customers for this service. They were also the first to develop an effective insulation for the cables using gutta-percha, a plastic-like material extracted from trees in British-controlled Malaysia.[23] To illustrate Britain's dominance, by the end of the nineteenth century, British companies owned 80 per cent of cable-laying ships and two-thirds of the world's cables.[24] After declaring war on Germany in 1914, Britain's first action was to cut the primary cables linking Germany with the rest of the world, forcing them to communicate by wireless.[25] This meant that while Britain's communications continued almost completely uninterrupted during the war, they could listen in to Germany's over the air.

As the international system of cables developed, networks were designed for security by ensuring a diversity of connections so that there was no single point of failure in the system. While most areas are connected by multiple cables, in 2011 an elderly Georgian woman cut off Internet access to all of Armenia for five hours when she accidentally cut a fibre-optic cable while she was scavenging for copper.[26]

The copper wires of the late-nineteenth and early-twentieth centuries were later replaced by analogue coaxial cables in the 1950s, made possible through the development of submersible repeaters, which would allow the cable's signal to be amplified under the sea. Cables could now carry telephone conversations and had a larger bandwidth. It also meant that ships and stations had to be redesigned to carry the repeaters. During the Cold War, the US military prioritised coaxial cables due to their stability and security as a means of transmitting secret information. The first deepwater cables with repeaters were laid between Florida and Cuba in 1950, forming a link that would soon connect the United States government to their puppet regime, led by the military dictator Fulgencio Batista.

The final technological shift that made the transportation of current Internet traffic and AI workloads possible was the development of fibre-optic cable and the laying of new networks in the late 1980s to early 1990s. The large conglomerates and nationalised telecommunication companies that had dominated the mid-twentieth century networks had become deeply intertwined with national security considerations. But in the political culture of the 1990s, deregulation and privatisation took over and the first fibre-optic networks were built by a consortium of private international companies.[27] In 1989, AT&T laid the first trans-Pacific fibre-optic cable, TRANSPAC-3, which followed the same route from the mainland US through Hawaii and Guam to Japan as the original Cold War-era cable laid in 1964.

Most cables in the twentieth century were constructed across the Atlantic Ocean, connecting the United States and Europe. It was only in the 1990s that countries surrounding the Pacific Ocean received greater connectivity in response to the growing importance of Asian markets in the global economy. In the 2000s, this was extended to parts of eastern and southern Africa through SEACOM, a new cable owned primarily by African investors.[28] There was a period of intense infrastructural development in the early 2000s, which corresponded with the early growth of the Internet. With more services connecting South and East Asia, Egypt now plays an oversized role in facilitating this global network, with a large part of the Internet's backbone passing from Europe to a range of other Pacific countries through a 'digital Suez canal' which takes advantage of Egypt's geopolitical position.[29]

While the first era of fibre-optic cables was owned by large telecommunication consortiums, in the late 2010s large tech companies began to finance their own submarine cables to exert more control over the basic infrastructure of the Internet. Google, Meta, Amazon and Microsoft have all invested heavily in their own networks. Google began to build its own cables in 2018 and now owns or has invested in nineteen of them including Curie, a cable connecting Chile and California named after physicist and chemist Marie Curie. Meta commissioned NEC Corporation to build the world's highest capacity submarine cable, a transatlantic connection between North America and Europe that will carry 500 terabits of data per second.[30] Today, most new cables are laid with the financial support of a Big Tech firm. This move by tech companies into the submarine cable market is part of their shift into data infrastructure and points to their growing power over our digital lives.

Infrastructural Power

The AI boom is driving increased demand for and investment in the data centre industry. Operating AI models requires ever-increasing amounts of computational power – or compute – which is the processing resources required to run applications. The bigger these AI models become, the more compute they require. Large language models (LLMs) have grown from 117 million parameters with GPT-1 in 2018 to 1.76 trillion for GPT-4 in 2023.[31] This increase has been enabled by more advanced chips and the ability to use parallel processing, in which multiple data processing tasks occur through chips operating concurrently. To meet the challenge of training these large models, AI infrastructure has needed to scale up to incorporate larger data centres, more chips and more powerful supercomputers. As a result of these growing needs, today's AI companies spend more than 80 per cent of their total capital on compute resources.[32]

The growth of computational needs has given rise to the expansion of data centres and the creation of enormous hyperscale cloud computing providers, sometimes called 'hyperscalers'. These are huge facilities that have the ability to scale large systems of memory, compute and storage. Providers of these facilities rent out compute resources accessed remotely to clients and can massively scale up and down as needed when increased demand is placed on their systems. This uses a lot of energy and places large demands on the local electricity grid, with a hyperscaler using as much power as up to 80,000 American households.[33]

Over half of these hyperscalers are based in the US, with 16 per cent in Europe and 15 per cent in China.[34] This offers an indication of the relative power and significance of these different geographies in the development of AI. Since 2015, the number of hyperscale data centres has more than doubled, with the total figure expected to pass 1,000 centres worldwide by the end of 2024.[35] During this period, there has been a broad shift away from on-premise enterprise data centres and towards companies leasing compute from hyperscale owners. Synergy Research Group estimates that hyperscale capacity will almost triple in the next six years.[36] Their analysis showed the global hyperscale data centre market was worth $80 billion in 2022 and is estimated to reach $935 billion by 2032.

The large tech companies are consolidating their power and remaining at the forefront of research on AI through their ownership over this pivotal infrastructure. Compute is currently monopolised by a handful of players that own the infrastructure necessary to power large AI systems. At the beginning of the platform era, following the rise of social media networks and marketplace platforms, the most critical part of digital businesses for generating value was in software. The companies that soared in value during this period were those that owned software platforms that could amass billions of users. Today, it is the infrastructural dimension of tech companies that is rising in prominence. While many companies have begun developing products and applications based on LLMs, only a select few own the computing hardware that enables these powerful models to operate.

This gives rise to the importance of Big AI's development of *infrastructural power* – a capacity to deploy large-scale computational resources for achieving strategic goals. The companies that have spent the last decade investing in such resources now have a competitive advantage over their rivals because they own the in-demand facilities that AI companies require to train and deploy their models. The capacity of many data centres is booked out months in advance and the market is experiencing a significant boost with the explosion of interest in AI. This concentration of infrastructural power in the US and with American companies also helps demonstrate why they continue to exercise the most power in the networks of AI. At nearly 300 centres, northern Virginia alone has more data centres than most countries.[37] Infrastructure is an important and indispensable part of the extraction machine; unskilled labour is cheap and can be located anywhere, whereas compute resources are scarce and tightly controlled by the industry's largest players. As we will see, this creates strategic tensions between the private companies that own the majority of these resources and the states that wish to direct them for geopolitical goals. At times, the objectives of companies can align with their host states, but AI companies are likely to prioritise their own commercial interests over the public good.

In 2021, just three companies – Amazon, Microsoft and Alphabet (Google's parent company) – owned over half of all major data centres.[38] In the first two quarters of 2023, these three companies accounted for two-thirds of all investment in the worldwide cloud computing market. With Meta attempting to catch up, just a few large tech companies currently account for a significant portion of growth in the sector. At the time of writing, Meta anticipated that capital expenditure would be between $30 and $37 billion in 2024.[39] Alphabet's capital expenditures totalled $11 billion in the fourth quarter of 2023 and were largely due to increased investment in AI infrastructure, servers and data centres.[40] Microsoft invested $10 billion at the start of 2023 in OpenAI and announced that it had an aggressive spending plan to build new data centres to support AI, with expectations of capital expenditures rising for each quarter of 2024. Amazon continues to announce large capital investments of $7.8 billion by 2030 to expand data centre operations in Ohio, $35 billion by 2040 in Virginia, and $13 billion in Melbourne and Sydney, to name just a few of their planned projects. China is also planning to increase its investment in AI hardware with over $15 billion earmarked for new data centres by 2026.

With such a concentration of infrastructure in so few hands, and with the infrastructure being essential for AI systems, AI becomes a utility that providers rent out for individuals and companies to access. Hyperscalers own the new means of production for the era of AI. Companies interested in developing any services based on AI capabilities will have to do so under terms and conditions set by the large companies and paying whatever fees are set. Of course, in addition to the hyperscalers, there is a market in which smaller companies rent out compute as a service on AI marketplaces such as Vast.ai, Render and Kudos, but these are all relatively small in comparison to what the large companies can offer. While it is still unclear who will be the major winners of the AI revolution, it is highly likely that big companies with a monopoly over renting 'AI-as-a-service' will benefit the most from whatever changes are coming. Just as the original decentralising ethos of the early web gave way to the centralised control of a handful of platform gatekeepers, the race for AI promises to further centralise power into an even smaller collection of large tech companies that own the infrastructure that makes it all possible.

The major bottleneck in the further development of AI is not necessarily in the availability of data, but in the computational power of AI infrastructure and in companies' ability to attract top-level talent.[41] These final two are closely connected. There is only a select number of people with industry-leading expertise in machine learning and these people tend to work for the top tech companies at which research and development of the fundamentals of AI is currently taking place. Jobs pay extremely well and workers can negotiate conditions with their employers. Competition over the leading talent is fierce, with three of Apple's star machine learning engineers reportedly moving to work on Google's AI projects between October and November 2022.[42] In addition, a number of younger startups such as Jasper, HuggingFace, Anthropic, Stability and Midjourney have also reportedly poached top players from leading tech companies, particularly from Google and Meta.[43] Despite this, Big Tech firms continue to dominate the race for AI talent, with the five largest hirers – Google, Meta, Amazon, Microsoft and Apple – collectively hiring ten times more AI talent than the next five largest hirers combined.[44]

Data centres are a key component of AI understood as an extraction machine. They are the engines of the machine that convert electricity, water, labour, critical minerals and data into the computational power for operating AI models. Constructing new data centres to accommodate the ever-expanding needs of the AI market requires more concrete, aluminium and steel to build the centres, and copper, lithium and silicon for the racks of computer servers. This requires large capital expenditure from companies and sets in motion a broader network of mining and processing minerals for the new sites. In the era of AI, hyperscalers wield a new form of power within the global economy. Data centres have always been an important part of tech companies' portfolios and the development of the Internet. They have tended to sit in the background, quietly ensuring the smooth functioning of the software we rely on every day. The expansion of basic tech services such as email, online shopping and social networks has required a rapid expansion of data centre capacity over the past two decades. With the rise of AI, hyperscalers now own an important asset that AI companies need to train foundation models. This allows legacy tech companies to monopolise this service and create strategic partnerships with AI startups they hope will offer them a competitive advantage over their rivals.

Will Google Drink My Town's Water?

As hyperscale data centres continue to grow, they have given rise to a contentious politics surrounding their consumption of electricity and water in addition to their employment practices. Any new data centre will draw significant resources from the local electricity grid and, depending on the location and the cooling system installed, will likely also require large amounts of water. Both these critical resources required to expand AI will potentially be in short supply in the coming decades. For example, by 2028, data centres in Ireland are estimated to require roughly 27 per cent of the country's available electricity supply.[45] 'Press Pause on Data Centres' is a campaign run by the anti-fossil fuel group Not Here Not Anywhere (NHNA) arguing there should be a moratorium on the construction of new data centres until they can be run entirely on renewables and that a cap should be set based on the state's climate commitments.[46] In 2022, Ireland's state-owned electricity provider EirGrid announced it would no longer be accepting applications for new data centres in the Dublin region until possibly 2028. Although not a full moratorium, and not including data centres already in the pipeline, this shows there is mounting pressure, particularly in the case of Ireland, where EirGrid has warned the country could face rolling blackouts as a result of a lack of electricity supply.[47]

Data centres in warmer climates are often also in need of a drink. A large data centre can consume between 11 and 19 million litres of water each day, equivalent to an American town of 50,000 people.[48] Google made headlines in 2021 when it publicly disclosed that it used 355 million gallons of an Oregon city's water in one year, which was 29 per cent of the city's total water consumption.[49] In July 2023, another Google data centre in Uruguay was in the spotlight because, as the country was suffering its worst drought in seventy-four years, protestors argued that the government was giving priority to multinational tech companies with enormous water needs.[50] In Google's 2023 environmental report, it stated it consumed 5.6 billion gallons of water in 2022, which was a 20 per cent increase on the previous year.[51] The growth of compute-intensive AI products is likely to make this issue much worse. (Although data centre operators have promised to bring down the total water used by centres, even as environmental protestors struggle to secure communities' water rights.)

In addition to environmental concerns, labour rights have also been a major issue. Data centres are often touted by tech companies as high-skilled job creators for the towns that host them. But most of them do not require a large staff. Beyond the design and construction of the building itself, the centres require a small number of technicians, facility managers, security guards, cleaners and other administrative staff. The quality of these jobs also depends on the hiring practices of the companies. Some workers are employed on permanent, full-time contracts as employees of the company, but many of the data centres of the largest companies are staffed by temps, vendors and contractors (TVCs) who do the same work as full-time employees of the company but receive less pay, are kept on short contracts and are stripped of other employment benefits. While these tech workers may not have it as tough as the data annotators – and, in many respects, their situations are vastly different – they have rightly fought back against being treated as second-class citizens at major tech companies.

In 2021, the magazine *Data Center Dynamics* revealed that Google's data centres employed an estimated 130,000–150,000 TVCs, a much higher number than full-time Google employees working in these centres.[52] Many TVCs are provided with three-month contracts, meaning that much like the data annotators in Uganda, they're left constantly worried as to whether their contract will be renewed. Contractors are also given a two-year maximum time limit for their work at Google, after which they have to wait six months to apply for a new position. This results in a two-tier system in which TVCs are treated with less respect than full-time employees and have fewer opportunities to progress within the company.

As we will see in Chapter 7, this greatly complicates the type of labour activism that is possible for both full-time employees and TVCs at Google. *Data Center Dynamics* reporter Sebastian Moss notes, 'It's a period of excess for the [data centre] sector – money is pouring in. But then they somehow can't afford to pay their workers properly? To have this clear disparity between an incredibly profitable company that promotes itself as a great place to work and the reality, which is that most of these workers are underpaid and overworked, it's grim.' When asked about Google's actions after the piece, Moss states, 'they have not done a thing, the only thing they said in response was, have you looked at Microsoft and Amazon?'

The AI Arms Race

As rival nations seek to harness AI's potential and become leaders in the development of this new technology, building AI will have important implications for geopolitics. Investment in AI within the United States and China dwarfs that of all other countries combined.[53] The governments of each country have made large long-term investments in AI, and see control of this technology as pivotal to expanding their influence throughout the century. As Kai-Fu Lee, chairman and CEO of Sinovation Ventures and former president of Google China notes, 'in this race there are not three medals like in the Olympics. There are only two medals and they belong to the US and China'.[54]

Maintaining a lead over other countries requires the development of multiple aspects of the AI value chain, from mining critical minerals and building data centres to establishing AI institutes and attracting top talent. In the case of critical minerals, this is the point at which debates around AI intersect with new forms of green energy, because everything from data centres to machine learning models will require renewable sources of power which rely on these minerals. Countries are racing to secure access and are investing in new infrastructure and projects across the globe. While some of these resources are based in Western countries – Australia has almost half of global lithium supplies – most are based in the Global South, which will almost certainly increase extraction activities for these regions, raising a host of questions about labour rights, environmental issues and exploitative trade agreements.

The growing importance of AI shifts attention from one set of resources used in abundance throughout the twentieth century to another. Critical minerals such as copper, cobalt, lithium and nickel will fuel our clean energy future and produce many of the tech products that will be essential in the era of AI. Many forecasters are already predicting potential disruptions to supply chains as demand for these minerals skyrockets. In the past five years alone, demand for critical minerals has doubled and could increase 3.5 times more by 2030.[55] Some are only found in great quantities in a select number of countries: most cobalt is mined in the Democratic Republic of Congo and most nickel comes from Indonesia, for example. There are concerns that even minerals we use in smaller quantities could become key bottlenecks in cases where they are monopolised by a small group of countries. China accounts for 98.1 per cent of the production of gallium, an in-demand material for semiconductor manufacture. Although this only amounted to 440 metric tons globally in 2022, China's announcement of export restrictions in response to US export controls on semiconductors hints at the kind of mineral trade wars that are likely to come in the next decade.[56] It is not just the sites of extraction, but where these minerals are processed that matters for global politics. A report by the International Energy Agency shows that China dominates the majority of the processing and refining of critical minerals, including 100 per cent of graphite, 90 per cent of rare earths, 74 per cent of cobalt, and 65 per cent of lithium.[57]

Among other things, these critical minerals are used to produce the top-end computer chips that power AI. In May 2023, US chipmaker Nvidia reached a $1 trillion valuation due to a surge in demand for their products following the boom of generative AI (and reached a $2 trillion valuation in March 2024). According to the 2023 'State of AI Report', Nvidia's chips are used nineteen times more in AI research papers than all other alternative chips combined.[58] Nvidia holds roughly 95 per cent of the GPU market for machine learning applications.[59] Selected Nvidia partners such as CoreWeave and Lambda make tens of thousands of GPUs available to rent and often sell out instantly when they place new capacity on the market. A handful of companies dominate chip production, with many of them based in Taiwan, Japan, South Korea, Germany, the Netherlands and the US.[60] The most prominent example is Taiwanese firm TSMC, manufacturer of Nvidia's renowned hardware, and the only firm able to manufacture the most advanced AI chips.[61] In September 2023, the company announced that for the next eighteen months, it would not be able to meet demand as the process of finishing the chips with advanced packaging was being slowed by the need to upgrade its facilities.[62]

Gulf states such as Saudi Arabia and the United Arab Emirates are also buying up chips in their efforts to become global leaders in AI. The Saudi government bought over 3,000 of Nvidia's H100 chips through the public research institution King Abdullah University of Science and Technology.[63] Meanwhile, the UAE has secured access to thousands of advanced chips and has developed its own LLM called Falcon.

The scarcity of AI chips has also fed competition between the US and China, with the Biden administration announcing a ban on the export of the latest AI chips to China in October 2022, in an attempt to hinder the development of its AI capacities. China's rapid advance in leading technologies and its control over key critical minerals and hardware has raised concerns in the US over its national security and economic competitiveness. In early 2023, a US State Department-funded study made international headlines when it found that China

leads in thirty-seven out of forty-four critical and emerging technologies, as the US and other Western powers failed to match China's
research output.[64] The report argued China had established a 'stunning
lead in high-impact research' and called for the US to 'rapidly pursue
a strategic critical technology step-up'. AI is considered by both sides
to be a pivotal technology that will provide them with an edge over
rivals in their struggle for global hegemony.

In the US, concern over AI is fuelling aggressive and militaristic
rhetoric. In May 2023, former United States Trade Representative
Robert Lighthizer told a House Select Committee that 'China is the
most dangerous threat that we face as a nation. Indeed, it may be
the most perilous adversary we have ever had . . . China believes it
is destined to be the world's only superpower and we are in the way'.[65]
In the same hearing, former Google CEO Eric Schmidt argued that
'It's never too late to stop digging our own graves . . . The technology
competition between China and the US is the defining moment of all
of the competitions. We must organise around "innovation power":
the way you win is to innovate ahead of the competition'. Schmidt
called for an enormous increase in non-defence R&D spending from
$2 billion to $32 billion by 2026. In another hearing of the same
Select Committee, congressman Mike Gallagher noted that 'we won
the Cold War in part by controlling cutting-edge technology . . . [but]
the Chinese Communist Party has learned from the Soviets' mistakes
and is pursuing global dominance of critical technology'. The US's
position as the global hegemon affords the country colossal economic
benefits (mostly at the expense of much poorer countries) and it is
not a role the political elite in the country are willing to give up
without a fight.

In the race for compute, states imagine AI as a tool they can control
to outmanoeuvre their rivals. But governments are increasingly reliant
on private companies. States announce hundreds of millions of dollars
of investment in AI, which pales in comparison to the tens of billions
invested in infrastructure by large tech companies. Hyperscalers have
become so powerful that they interact with and influence national
governments on AI industrial policy. Big tech players in the world of
AI are invited to the White House and produce policy ideas for the

regulation of AI, leading to a small handful of companies playing an oversized role in the debate. In July 2023, Amazon, Anthropic, Google, Microsoft, Meta and OpenAI signed an agreement with the White House to have new guardrails in their generative-AI content to enhance its safety and security.[66] Four of these companies also formed the Frontier Model Forum, 'an industry body focused on ensuring safe and responsible development of frontier AI models'.[67] The idea is to pool resources to actively guide the development of AI policy to the benefit of the biggest actors in the market.

AI has become the focal point of competing ideas for how our future will develop. Corporations see dollar signs; states see military hardware and competitive economic advantage. What seems unlikely is that these interests will lead to AI being developed in ways that benefit humanity. Very rarely are users themselves asked what kind of AI they would like to see developed, particularly those from marginalised communities who are most likely to be affected by AI's social harms.

The networks that constitute AI are vast and encompass a growing number of nodes. Even a country as small as Iceland, and a data centre with only three permanent staff such as Einar's, can be connected to a broader system of AI development. The construction of fibre-optic cables that connect most of the world enables different nodes of this network to communicate in milliseconds. These connections enable new economic opportunities but also draw actors into complex systems of power that dictate the terms on which they can participate in the network. The infrastructural power exercised by large tech companies in this network enables them to determine the conditions under which smaller parties must negotiate with them. They can use this power to lock in strategic partnerships, charge more for the use of their services and maintain control over who has access to the most cutting-edge technology.

For Iceland, the economic benefits of capturing a larger share of the data centre market are immense. But despite all their efforts, these countries are likely to remain marginal players in the larger global race for AI. This is not really a problem for Einar and the Icelandic data centre industry, because there are limits to the amount of new capacity their national power grid could accommodate. The industry has brokered a political deal with the government to facilitate a managed rate of growth over the next decade, but a large spike in capacity would place too great a burden on national energy resources. AI has offered new opportunities for Iceland to continue a steady growth in capacity and take on certain non-latency-sensitive workloads of AI companies.

This chapter has been focused on one of the edges of the AI global production network, thousands of kilometres away from one of its centres in California, where important decisions about investment and product development are made. In the next chapter, we turn to another important node in the network: the creatives whose work is used to train generative AI models, who are faced with the ever-looming potential that they are training their own replacements.

4

The Artist

Laura would never forget the moment she first realised. Her sound engineer friend called her to ask about her experience of selling an AI company the rights to her voice. At first, she thought her friend was confused. Laura was a professional voice actor, but she had never worked with an AI company. She got into voice acting after taking a class at drama school, and her work has since appeared in advertising, cartoons, computer games and audiobooks, alongside her acting and writing work in theatre and television. It's hard-won experience built up over nearly twenty years.

She asked her friend where he found her voice online and if he could send her the link. A moment later, she opened the website and saw an avatar of an Asian woman named Chloe. She was advertised as having a neutral Irish accent that was great for audiobooks. Laura clicked on the image and began to hear her own voice: 'Hello, my dear ones, my name is Chloe. I have a soft and caring voice, I can record voiceovers for audiobooks and educational videos, and even make a soft sell. I can do any voiceover that you need.' It was a little deeper, quite robotic, but unmistakably her voice. She was in shock and disbelief. She had no idea what this company was or how it had created a copy of her voice. However they did it, Laura realised she was now competing for jobs with a synthetic version of herself.

It took a while for everything to sink in. How had this company even captured her voice to clone it? She spoke with a media lawyer friend and did some research into the company. She knew that under EU law a voice itself could not be copyrighted, but that she could contest the company's use of her voice and claim performance rights. Her friend said that her rights would depend on where the company was based. This is where things got sus. The company didn't list a business address on its website or even indicate which country it was based in. She emailed them with an innocuous inquiry and stated that she needed their address. The reply came back: they would not disclose an address and could not assist her further with her inquiry.

She was furious. Was there nothing that could be done? Voice actors like herself would be utterly defenceless against companies stealing their voices and creating digital copies that would put them out of business. This was theft, pure and simple. Her voice had been taken without her consent and was used to create a robotic monster that could be bought for one tenth of the price of a human actor. It was as if somebody had cloned her and produced a Frankenstein-like soulless twin.

After some more sleuthing, Laura discovered the source of the problem. She had done a job for one of the big tech companies a few years back and the terms of the lengthy contract contained some clauses she hadn't focused on – simply put, the company owned the rights to her voice 'in perpetuity'. As it turned out, the tech company had then sold the recordings to a third-party AI company that was legally entitled to produce the synthetic clone. At the time of signing the contract, she was scarcely aware this was even a possibility. Such contractual terms stipulating actors must sell their rights for ever are blatantly unfair, but well-paid jobs can be hard to find and many voice actors seeking work are forced to sign whatever terms are put in front of them. Individual workers facing a pandemic or other hardships are not well-positioned to play hardball with multinationals.

She spoke with a lawyer about her issue and it seemed nothing could be done. While quite possibly unethical, there was nothing illegal about the actions of the tech company or the AI company. It felt like the law had not caught up with the technology and that her entire industry was now even more vulnerable to exploitation. Many companies with tight budgets would begin using AI clones at a fraction of the cost of voice actors. Millions would be at risk of losing their jobs and their identities as actors. The technology was still nowhere near the level of a professional actor, but for many this would not matter.

As she listened to Chloe yet another time, she couldn't help but think that something important had been lost in the transformation. A human voice has something to it that you can connect with. Humans use specific intonations and inflections at key moments to evoke emotions and meaning that is difficult for AI to replicate. What will happen to her industry – and to society – when the world is full of these dull and lifeless creations? Are her days as a voice actor numbered? Will the technology eventually become indistinguishable from a real human voice? Only time will tell. For now, Laura has to resign herself to having one more competitor in the industry.

Art Without the Artist

At the end of 2022, following the launch of new tools such as ChatGPT, Dall-E and Midjourney, a wave of AI-generated art swept across the Internet. Do you want the AI image generator Stable Diffusion to create a painting of Kermit the Frog in the style of Francisco Goya's black paintings? You got it. A new Oasis album that sounds like their 1995–97 period? Done. A poem by T. S. Eliot riffing off lyrics from Megan Thee Stallion? ChatGPT will give it a go. Disney used AI to create the credit sequence for the Marvel series *Secret Invasion*, while Netflix Japan went further, announcing that one of its latest animated shorts was produced by generative AI.[1] The Writers Guild of America has concerns that studios will begin using AI screenwriting tools to produce the basis of movie scripts and only hire humans to edit and polish them. Artists and creatives of all stripes are in a pitched battle to control the use of this new technology that threatens to steal their identities, devalue their work, and potentially replace them.

Although often without adequate recognition, generative AI stands on the shoulders of giants. The source of its sometimes remarkable abilities is the enormous corpus of human data on which it is trained. For large language models, recent advances have been possible due to increases in the size of their datasets, with current models using hundreds of billions of parameters. As these datasets are so immense, it is highly likely they contain copyrighted material. Some authors have complained that models like ChatGPT were trained on datasets containing numerous copyrighted works obtained from 'shadow libraries' such as Library Genesis and Z-Library, which unlawfully distribute books online.[2]

The capabilities of AI image generators are based on similar techniques. Stability AI trained its AI model Stable Diffusion on a large dataset called LAION-5B, an open, publicly available dataset of 5.8 billion image–text pairs scraped from the Internet.[3] Artists have used tools like ihavebeentrained.com to detect whether their work has been included in the dataset and found that it contains many artists' images used without their consent. A trio of artists brought a lawsuit against Stability AI and Midjourney, two AI companies that create image generation tools, alleging the AI tools have infringed their rights.[4] David Holz, the founder of Midjourney, admitted to not seeking consent from living artists for work still under copyright before using them to train Midjourney's model.[5]

AI companies are essentially selling what they don't own: the value of image generator tools is based on the human originals, ripped from large datasets and monetised on the sly. These companies have adapted the tech industry's famous motto 'move fast and break things' to 'move fast and steal things', while giving little consideration to the human creators who will be most affected. The image generators allow anyone to knock off an artist's style and threaten their livelihoods without any remuneration or credit for using their work. People have been making fakes and derivative works for generations, but AI poses a completely new threat because millions of users can now do this at the click of a button. Rather than selling single imposters, AI tools loot en masse, ransacking an entire community for its valuable creative ideas.

The struggle to protect artists from other people profiteering off their work did not begin with AI. For a long time, artists have been fighting unfair terms in their contracts and being asked to sign away their rights to future revenue from their work. One voice actor we interviewed as part of our fieldwork informed us that royalties for the number of works sold used to be more common, but a flat fee no matter how many times the work was sold was becoming the industry standard. 'The issue of the creator being the least important person in the value chain and everyone else being on top of them has been around for ever. AI hasn't created the problem of exploitation, it has just exacerbated it,' she noted.

But there are some people who are already trying to hack the system. Visual artists now have a new tool, developed by researchers at the University of Chicago, to help them fight back. The tool, called Nightshade, enables artists to add special pixels into their digital images that, if used in an AI training set, could cause the model to break down.[6] This is just one of the small ways in which artists have begun trying to resist this new form of exploitation.

With every new model, AI tools become increasingly more sophisticated and adept at mimicking their human creators. Early examples of generative AI images struggled to produce detailed human faces and depict the correct number of digits on hands, but such constraints were quickly surpassed. Chatbots are now great at standardised tasks, yet are still unable to read between the lines. ChatGPT-4 can pass the Uniform Bar Exam, a law exam in the United States, but still produces outlandish hallucinations and fails to correctly answer simple questions requiring contextual understanding. No doubt, even these limitations will rapidly be overcome.

For voice actors, there is still a sizeable gap between the performance that can be achieved from a real human being's voice and that of AI. For the time being, professional high-quality recordings for big brands will still need human voice actors. The problem is almost everyone else will choose the cheaper option, and these are the jobs that help sustain many people's careers. On one AI voice website, subscribers can access a catalogue of hundreds of voices that can express various accents and emotions through text-to-speech software for $27 a month. This cost is tiny when compared to hiring a human and can be done in seconds. As the technology improves and we all become accustomed to listening to AI voices, jobs for human actors may become harder to find.

Generative AI is also likely to transform the entertainment industry. In the United States, the right of publicity protects artists against the exploitation of their identity, although the precise scope of this protection varies state by state.[7] In other common-law jurisdictions such as Australia and Canada, this right generally falls within the scope of the tort of 'passing off', which prevents others from exploiting one's image or likeness, while most civil-law jurisdictions such as France and Germany have specific provisions that protect an individual's image. The problem is that AI creates new ways in which people's image and likeness can be exploited by companies. A performer could sign on to one season of a show only to have an AI-generated version of themselves perform in the following five seasons. Symbolic of the shift, James Earl Jones officially retired his Darth Vader voice, but not before selling the rights to his voice to a Ukrainian tech company, which will use archival footage to create new dialogue. Resurrecting artists is not new – *Star Wars* brought Carrie Fisher back after her passing for *The Rise of Skywalker* – but generative AI promises to expand this even further across the industry. The problem is not necessarily with the technology per se, but with how it is used. It's not hard to imagine a future in which all artists are asked to sign contracts allowing their talents to be synthesised and used in whichever way the company wants. Some actors are already being told they will not be hired without agreeing to new AI clauses, while others are discovering them deceptively embedded within their contracts in confusing and ambiguous language.

We shouldn't over-inflate the capacities of our current crop of generative AI products. People using image generators are still circumscribed by the limitations of the prompt input. Chatbots can impress in specific tasks, but usually lack the flair and creativity of genuinely interesting writing. AI voices can communicate in different registers, but not with the depth and complexity of a human voice actor. Many video game developers remain sceptical of the potential of AI to play a significant role in creating games on the scale of *Cyberpunk 2077* or *Red Dead Redemption 2*.[8] There is also very little appetite for AI voices from gamers. For the amount of money companies charge for video games, consumers expect excellent quality that will keep them immersed in the gaming experience. Although this might change, for the foreseeable future, there will still be a premium on human-generated creative outputs, particularly for high-profile and big-budget projects.

However, there are different tiers of work in creative industries and many of the smaller jobs that are the bread and butter of certain artists will disappear. These are the jobs that even successful artists require to make a living and their disappearance will likely devastate the ecosystem within which artists can eke out a career. One artist we interviewed mentioned, 'there's a lot of day-to-day stuff that's not glamorous, but it pays the bills. It's these jobs that may fall away. I record dictionaries, English learning materials, explainer videos, and corporate videos. I enjoy it and it's how I can get by to do the other stuff.'

The disappearance of these jobs could also have a big impact on the types of people who can become professional artists. For the most part, art is already poorly paid. Eliminating a broad range of decent jobs from circulation could mean that the only people able to devote themselves to the arts are the children of wealthy families. In a 2018 study of visual artists, nearly half of the those surveyed attributed less than 10 per cent of their income to their art practice, with nearly one in three relying on family support or inheritance.[9] AI will take away a lot of the work that sustains a career in the arts. For example, AI-generated images will certainly challenge traditional image libraries, through which many commercial artists make a living. AI could also produce work for decorating homes and offices and adorning book and album covers. New technology is often claimed to create new jobs for those it destroys, but for artists, this will come as little comfort as these jobs will primarily be in tech and will not require the special-isation artists have developed through years of practice.

Some artists are fighting back and winning against their likenesses being stripped and packaged for generative AI. The use of AI in Hollywood became a central point of contention in negotiations as the Writers Guild of America – representing 11,500 screenwriters – and the Screen Actors Guild-American Federation of Television and Radio Artists (SAG-AFTRA) – which represents 160,000 performers and media professionals – went on strike in 2023 as part of a broader dispute over working conditions in Hollywood studios.

Screenwriters were concerned that multiple studios had plans to use AI to generate scripts based on books and other works in the public domain. The Writers Guild of America argued that AI should not be used to create scripts and that its members should be paid in full for all work related to scriptwriting. In September 2023, the Writers Guild struck a deal with the studios that stipulated AI cannot be used to write or rewrite any scripts and which protects writers from having their scripts used to secretly train AI scriptwriting bots.[10] This was a huge win and a potentially precedent-setting deal that was likely the first time enforceable terms that protect scriptwriters were put into an agreement placing strict limits on how AI could be used.

In their contract negotiations, SAG-AFTRA focused attention on protecting its members from losing income through the 'unregulated use of generative AI'. President of SAG-AFTRA Fran Drescher stated that 'artificial intelligence poses an existential threat to creative professions, and all actors and performers deserve contract language that protects them from having their identity and talent exploited without consent and pay'.[11] During the dispute, SAG-AFTRA claimed that Hollywood executives had put forward a proposal that background performers could receive one day's pay to scan their likeness and use it in perpetuity with no additional compensation.[12] In December 2023, the studios and the union agreed to terms that would govern the use of generative artificial intelligence.[13] While this was a historic win for the union, it does contain certain ambiguities that may prove difficult to interpret in future disputes. The deal mentions 'synthetic performers' created by generative AI and states that if a 'Synthetic Performer [has] a principal facial feature (i.e., eyes, nose, mouth and/or ears) that is recognizable as that of a specific natural performer' then the company must obtain that performer's consent. The question, then, is what is 'recognisable' as a feature of one specific actor? What if a company produced a Morgan Freeman-like voice that didn't exactly sound like him but conveyed a similar gravitas and was designed as a copycat? Actors are also concerned that the language is too weak on certain provisions and that there are too many loopholes that could be exploited by studios.[14]

Voice actors are particularly vulnerable to being replaced by AI and are among those leading the fight. Voice actors from across the globe have come together to form the United Voice Artists (UVA) coalition, which consists of over twenty voice acting guilds, associations and unions. They have launched the campaign 'Don't Steal Our Voices' in an effort to pressure governments into regulating the use of AI in the creative industries.[15] 'UVA calls on politicians and legislators to address the inherent risks, both legal and ethical, in the conception, training and marketing of AI-generated content,' states UVA in a press release; 'any use of AI technology to generate and clone human voices must be subject to the explicit consent of the voice-over artists and performers, which must therefore be in a position to refuse the use of their works and performances, past and future, for purposes not expressly authorised by them, and be offered practical solutions to ensure the effectiveness of this choice.'[16]

In the hands of Hollywood executives, AI becomes a weapon used in their ongoing attempts to control artists and their creative works. What has happened in Hollywood will likely play out in other creative industries, only some of these artists will have less collective power to negotiate favourable terms. And this is not to mention a wide range of other professions in which creative labour is required, such as teaching, architecture, product design and web development. Companies will be looking for ways to automate employees' or contractors' labour to pay them less. This struggle is only just beginning and will likely further evolve as the technology develops and artists continue to defend their practice. This struggle also raises the important question of whether AI could be called creative, and if it could be understood as similar to human intelligence and creativity.

The Creativity Test

AI may be able to perform some routine tasks traditionally done by creative professionals, but could AI possess genuine creativity? Two of the great thinkers of AI disagreed on this very question. Ada Lovelace was an English mathematician and writer of the nineteenth century who was one of the first to recognise the importance of Charles Babbage's machine, the Analytical Engine, often considered the first computer. She believed the machine capable of a great many things, but ultimately incapable of independent learning. Because computers merely executed commands, she considered that any cre-ativity should be attributed to programmers: 'The Analytical Engine has no pretensions whatever to *originate* anything. It can do *whatever we know how to order it* to perform'.[17] For Lovelace, the real limitation of machines was their inability to come up with something new outside of their programming. They merely executed tasks designed by their programmers.

Alan Turing, an English mathematician and computer scientist, who many consider the father of modern artificial intelligence, believed that computers could in fact surprise us, particularly where the consequences of certain states of affairs were not immediately recognisable to humans. Turing believed that a machine could be said to exhibit intelligent behaviour if it could pass what he called 'the imitation game', which is now better known as the Turing test. In a paper published in 1950, Turing proposed that if a human evaluator had a text-only conversation with a machine and could not reliably distinguish it from a real human then the machine would pass the test.[18] The purpose of the Turing test was to develop an alternative to the more ambiguous and potentially un-answerable question of 'can computers think?'. The problem with the alternative formulation is that passing the test turns on whether a programmer can trick a test subject. The emphasis is on the process of deception rather than the capacity of the machine or the value of its outputs. For example, a chatbot called Eugene Goostman developed in Saint Petersburg in 2001 by three programmers is often considered the first to have passed the Turing test.[19] Goostman was introduced to participants of the study as a thirteen-year-old Ukrainian boy, a background intended to excuse his poor language skills and lack of general knowledge. So when the chatbot responded to questions with strange non sequitur and nonsense answers, participants may have been encouraged to overlook these, potentially undermining the validity of the results. The imitation game simply encourages tricksters.

In the early 2000s, Professor of Computer Science Selmer Bringsjord and his team saw this problem and attempted to develop an alternative test. They wanted to verify whether AI could exercise human-like creativity, developing a new formulation which they named the Lovelace test.[20] The researchers proposed that AI could be said to be creative if its programmers could not account for how it produced an output. The point was to insist on a certain *epistemic* relationship between a human programmer and a system that produced outputs. In other words, did the programmer *know* how the machine did it?

And yet this alternative also proves unable to test whether a computer exercises genuine creativity. There are two main issues. The first is that computers may be able to produce outputs that their programmers cannot account for, but which might still be entirely lacking in worth or value. To focus exclusively on the epistemic relationship is to ignore the quality of what is being produced. For a computer to be counted as creative or intelligent its creative outputs must have artistic value for people and add something new to the world. As we encountered in Chapter 2, programmers of the current generation of chatbots already cannot account for certain outputs, particularly when they are led down hallucinatory paths. But we wouldn't count unintelligible outputs as instances of creative flair. The second problem cuts in the other direction: the test is too restrictive. Given enough time, developers of a program should be able to offer an account of the underlying reasons why it has been able to perform certain actions. The real question is not whether we can understand it, but whether it has produced something new that could be described as a valuable and creative work.

A better test, and one we hope still captures the original spirit of the Lovelace test, is what we call the Creativity Test. An artificial agent or system could be said to exercise genuine creativity if it can produce an output that originates something new that is judged as valuable by human observers. There are two elements to this test. The first is Lovelace's originating principle. Even if elements of a computer's output are present in its training data, its product must contain an original element that can be considered a novel contribution provided by the process undertaken by the computer. Secondly, this creative act must have value as an artistic creation. This is not to say that it must fetch a high price on the art market or that it must command universal adoration from critics, but there must be some sense in which its outputs are considered worthwhile by a human community. On both elements, there is a degree of subjectivity involved in any assessment, but we see this as unavoidable for any creativity test. Ultimately, the subjectivity of the test makes it difficult to use as a measurement tool for scoring specific creative systems.[21] It is rather intended as a thought experiment to help us understand the important elements involved in whether an AI system could be said to exercise creativity.

To determine how close AI currently comes to passing the Creativity Test, let us first consider how AI outputs could be compared to human creativity. The field of computational creativity is focused on modelling and understanding creativity using computers.[22] One of its objectives is to see if a computer could be capable of achieving human-level creativity. It is widely considered that humans possess a spontaneous capacity for innovation. We can come up with new ideas, insights and creative ways of understanding and representing our world. This creativity relies on a combination of bursts of insight and long periods of hard work developing our skills, often in conversation with others.

On the one hand, artists commonly describe some of their best work coming to them as a flash of brilliance. Many ancient philosophers used to attribute artistic greatness to divine inspiration. In Plato's early dialogue *Ion*, for example, he presents poetry as a result of divine madness, a form of truth revealed through the poet as a prophet of the gods. One version or another of this view still holds sway in modern times. Paul McCartney described waking up one day with a song playing in his head which he assumed must have been written by somebody else. After asking others if they had ever heard it before he stated, 'eventually it became like handing something in to the police. I thought if no one claimed it after a few weeks then I could have it.'[23] German polymath Johann Wolfgang von Goethe contemplated how to write *The Sorrows of Young Werther* for two years until suddenly it came to him: 'at that instant, the plan of Werther was found; the whole shot together from all directions, and became a solid mass, as the water in a vase, which is just at the freezing point, is changed by the slightest concussion into ice.'[24] Many of us could recount such experiences where ideas simply appeared to us.

On the other hand, such bursts of creativity could not be translated into great works of art without a long process of training and development. It is unlikely that had McCartney not received music lessons, played multiple instruments from an early age, and been among other creative minds, he would have written such transformative music. Many artistic breakthroughs that fundamentally change how a discipline operates are based on a thorough understanding and familiarity with a corpus of existing works. Creativity may appear to be a spontaneous activity, but the actual practice of creation often requires long stretches of artists honing their skills and messy periods of trial and error.

It is also important to note that there are many aspects of human thinking and creativity that could be described as algorithmic. The process through which we learn new skills is often rule-following and based on repetition and reinforcement. Many ideas that one might want to describe as innovation are often forms of imitation with subtle differences that distinguish them from that which has been copied.

While it is debatable whether AI has truly *originated* a new idea, it is clear that it can do more than that which Ada Lovelace thought possible. Deep artificial neural networks enable computers to generate far more surprising outputs from their input data than simple processes of command and execution. Contemporary approaches to machine learning allow computers to mimic the process of a child learning new patterns. One type of computer program that has surprised its creators is the modern chess engine, particularly those that use neural networks and reinforcement learning to master the game. An engine called AlphaZero, developed by Google DeepMind, became one of the strongest programs through a novel technique: rather than learning from previous games of grandmasters, AlphaZero developers only taught the program the rules of the game and nothing else, allowing it to play itself millions of times with the goal of achieving a better position until it attained superhuman performance.[25] As a result of this style of reinforcement learning, the engine discovered new moves that had never been played by humans, appearing completely counterintuitive based on how a human would play the game. While primarily based on logical calculations, chess has an aesthetic quality in which discovering certain moves relies on a strong imaginative capacity. Russian grandmaster Mikhail Botvinnik thought 'chess is the art which expresses the science of logic'. In this game and others, AI has proved itself capable of generating beautiful moves that surprised its creators and masters of the game. Some would argue that on our Creativity Test, a chess engine could be said to have passed. These engines have produced new ways of understanding the game that are considered valuable contributions at the highest level of play. Yet the objection could still be made: an innovative move in a rule-bound game is one thing, but what about a genuinely spontaneous and creative idea?

One thing is certain: AI does not create *ex nihilo*. AI's creations are based on its training data, from which it discovers patterns and produces outputs that resemble the data. In 2018, Christie's auction house announced its intention to sell the first piece of AI-generated art at auction.[26] *Edmond de Belamy* is a blurry portrait of a man based on a training set of 15,000 portraits from the fourteenth to the nineteenth centuries. It was produced by French art collective Obvious, based on a type of image generation called generative adversarial networks (GANs), invented by Ian Goodfellow in 2014.[27] While initially estimated at less than $10,000, the painting sold for $432,500 and achieved worldwide media attention for AI's ability to generate unique artworks practically indistinguishable from human art. GANs work by using two neural networks to generate and then judge the authenticity of an image until the AI can create convincing copies of its dataset. Although everything is ultimately based on its training data, these image generators can combine elements in new ways to produce strikingly original pictures.

The sceptic might still see copying and creating images – even if there is a degree of novelty – as fundamentally a reproductive act and not quite sufficient to pass the Creativity Test. Yes, the painting is new, but ultimately it is simply derivative of the other works in the dataset. But what about a novel? The creator of the Lovelace test, Selmer Bringsjord, stated that if an AI could write a novel that captured his attention and that he found compelling, this would satisfy his criteria.[28] In this field, AI still has some way to go. In 2016, *The Day a Computer Wrote a Novel* was celebrated as the first AI-generated novel and even passed the first stage of a literary award.[29] However, the team of developers behind the novel predetermined a large amount of the guidelines for the program themselves, including the plot line, characters and even key sentences and phrases, amounting to roughly 80 per cent of the novel, according to one of the developers.

Left to its own devices, AI struggles to produce a coherent piece of work. Another AI-generated project, the novel *1 The Road* (2018) sought to emulate Jack Kerouac's *On the Road* through an American cross-country road trip.[30] Writer and engineer Ross Goodwin drove from New York to New Orleans in a car equipped with a camera, microphone, GPS and a portable AI writing machine in an attempt to reproduce the experience of a road trip with the AI writing as they travelled. Here is the opening line of the novel: 'It was nine seventeen in the morning, and the house was heavy.' The results of this exercise were lacklustre, with most of the prose appearing nonsensical. However, these obvious limitations of AI for writing novels have not prevented people from using it to produce large quantities of AI-generated books to sell on Amazon, including books written under the names of famous authors without their consent.[31]

Whether or not the Creativity Test could be said to have been passed depends on a domain-specific assessment. There is a strong case, for example, that chess engines have created truly original and valuable moves – if one considers these creative acts. Image generators have also produced outputs with striking quality and originality. AI writing tools, on the other hand, are still much further from passing the test due to the low quality of their outputs and the amount of work that human programmers must inject into the process.

Ultimately, however, many studios in the creative industries are not concerned about whether AI art is genuinely creative or not. If it makes money and passes for a marketable product then it will likely be widely used. Art, too, is a commodity in a capitalist marketplace, and must be viewed from this perspective to understand how studios are likely to respond to new technologies. For the authors' part, we see glimpses of creativity in certain AI outputs and understand the reasoning of people who want to see these as examples of genuine creative expression. At the same time, we argue there are hard limits to what AI can produce when it comes to genuine works of art.

Why We Won't See a Computational Caravaggio

Computational creativity faces insurmountable problems attempting to emulate human art when it comes to some of the most important things that make us human.[32] Art is a cultural product that is made by humans living in communities and draws from their rich experiences, culture and history. Most artworks are produced through *intentional* activity that embodies a feeling or thought in an external form. The most fundamental issue with AI outputs is that they are not produced by a conscious entity that understands the meaning of its training data. Generative AI has no capacity to contemplate its own significance and respond in a way that brings something new into the world. Humans can reflect on their own condition and the world around them through creative processes that draw from a complex inner world of thoughts, desires and memories. An algorithm cannot read between the lines or take imaginative leaps from one concept to another. The creative process goes beyond that which can be programmed. Human beings will respond in dramatically different ways to stimuli that are difficult to predict. Two individuals living next to each other could become writers and compose completely different novels because of their unique experiences and personalities.

Creativity is also intimately connected to our material embodiment in the world. We are not minds in vats computing ideas and generating disembodied works of art. Human sense-making is rooted in our physical bodies and lived experience. We have bodily needs and a sense of our own precarity and finitude. Our consciousness of a realm of ideas and abstract concepts is mediated through our flesh, muscles and nerves. This physicality is an important part of our creativity and our bodies' capacity to learn artistic skills and improve our dexterity and coordination. Art reflects upon physical desires and pain. It can help us make sense of and cope with suffering, just as it can express the pleasures and joys of human experience. This material reality is the condition of possibility for art that reflects something important about human life.

One reason many artists are offended by the idea they could be replaced by AI is that this relies on an incredibly reductive view of what art is and how it is made. Musician and writer Nick Cave, for example, responded to the countless AI-generated songs 'in the style of Nick Cave' with the observation that ChatGPT has produced 'replication as travesty . . . a grotesque mockery of what it is to be human'.[33] Art is more than the ability to paint a pretty picture; it helps us understand ourselves and our place in the world. It is a form of self-expression that allows those around us to access our thoughts and feelings through an external form. The urge to define ourselves through creative works and to share these with others is part of what it means to be human. Some artists describe their work as a journey of self-discovery and a way of actualising themselves in the world. The novelist and poet Mary Ann Evans (known by her pen name George Eliot) considered 'art the nearest thing to life; it is a mode of amplifying experience and extending our contact with our fellow-men beyond the bounds of our personal lot'.[34] It is how we experience the extremes of human emotions and share these with others.

German philosopher Immanuel Kant, one of Western history's leading thinkers on aesthetics, considered that great art required human genius to produce it. AI could create what Kant called 'mechanical' or 'agreeable' art, which he separated from the skill and creativity required to produce true works of genius.[35] Such art could still be tasteful and aesthetically pleasing, but it would be fundamentally 'soulless' and lacking the qualities that would make it more than just a beautiful object. For Kant, great art requires a rational and conscious being to create a work so that it 'promotes the cultivation of the mental powers for sociable communication'.[36] The key to his definition is that art is fundamentally a *social activity* that takes place before a community of human observers. Great art could be shared with others and allow them to view the work and obtain a similar elevated yet harmonious mental state as the genius had when they created it. For both nature and art to be beautiful, for Kant, they need to be seen 'as if' they are the product of intelligent and purposive design. AI can produce rule-bound and aesthetically pleasing creative works, but these will always fall short of the intentionality and inspiration that Kant thought necessary for truly great art.

In a world flooded with AI art, we also completely lose art's relation to truth. Art is historically grounded and expresses something of a particular culture and people, with an artist drawing ideas and experiences from the society in which they live. In his writing on art, another German philosopher, Martin Heidegger, argued that great artworks help human societies understand their experience and define important moments and events. The work of art presents us with a coherent world in which certain things show up as meaningful and important. In one prominent example, Heidegger imagines how an Ancient Greek temple would have brought together and given meaning to that culture's ideas of birth and death, humans and gods, noble and base – reflecting their understanding of the proper way to live a human life.[37] In short, works of art can enable human beings to derive truths that give meaning and significance to their world.

It is also hard to imagine AI-generated art playing a political role as a critical and subversive tool, something that could speak truth to power like human-generated art. Art is able to ridicule and undermine authoritarian rulers and create new ways for people to see the world beyond the propaganda of the ruling classes. To perform this role, artists must be attuned to their political culture and be able to unmask unjust policies and inauthentic messages. George Orwell was motivated less by aesthetic considerations than the political power of his writing: 'When I sit down to write a book, I do not say to myself, "I am going to produce a work of art." I write it because there is some lie that I want to expose, some fact to which I want to draw attention, and my initial concern is to get a hearing.'[38] AI will struggle to replace the role of the critic and political dissident able to reflect on the subtleties of political life and tear down the comforting illusions of the wealthy and powerful.

Humans perform some tasks that are routine and could be completed by an algorithm, but they are also capable of so much more and can attain heights of creative excellence that artificial intelligence will never reach. We should not confuse mimicry with genuine human experience and conscious reflection. Unless there is a complete step-change in how AI learns and executes tasks, only humans can go beyond their programming.

The Curse of the New

What will society be like following the proliferation of AI-generated art? Will this new technology lead to the demise of other art forms and the neglect of human creativity? Will our society be fundamentally impoverished by this process, losing something important in the automation of creative outputs? Such concerns are not specific to AI and refer to what we call the 'curse of the new': oversized hopes and fears placed onto new creative technologies as they arrive on the scene. Writing in the early twentieth century, philosopher and cultural critic Walter Benjamin considered that mechanical reproduction of a work of art devalues its 'aura' and uniqueness.[39] During his time, mass-produced art in the form of film and photography was growing in popularity. He believed that a work of art derived its aura from the specificities of its unique historical location and origins. 'The uniqueness of a work of art is inseparable from its being embedded in the fabric of tradition,' he noted, which is why an original would always have a special status over a reproduction.[40] Human art has an authenticity to it because of its unique existence created by an artist at a particular moment in time. He predicted that as the reproduction of art became more widespread it would lead to a disenchantment with art and a loss of its meaning.

Yet are these concerns well-founded? Mechanical reproduction has existed in the art world in one form or another for centuries. Rembrandt operated a busy workshop in which he likely had a large number of apprentices working on his paintings, just as Michelangelo allowed some of his assistants to fill in parts of the Sistine Chapel. More recently, Andy Warhol operated an entire workshop in which the majority of artworks were produced by his assistants on a mass scale. Does the introduction of new tools necessarily undermine the possibility of art's role in our society? Technology has assisted artists in the act of creation ever since simple percussion instruments and cave paintings. Synthesised drumbeats and turntables did not destroy pop music, they merely enabled new genres of music to emerge.

When a new range of generative AI tools were released to the public in 2022, they became wildly popular as people experimented with new modes of creating. But so much of the excitement behind the current generation of AI tools is about their capacity to increase shareholder value. Companies are making bold claims about how transformative their products are and the artistic revolutions they will enable. Everyone is ready to jump on the AI bandwagon. But all new technologies go through such hype cycles with exaggerated hopes and fears about coming changes. Like other technologies that have come before it, AI is a tool that offers certain affordances and can be used for a variety of purposes. To put things in perspective, it's useful to compare the rise of consumer-friendly generative AI with the arrival of other era-defining creative technologies.

When photography first emerged in the mid-nineteenth century, it sparked fears that it would make painting obsolete. Photographers could capture the world with such precision and accuracy that it seemed to rival traditional forms of portraiture and landscape painting. People were divided on whether this should be feared or celebrated and on what impact this new technology would likely have on the art world. These debates were not dissimilar to concerns with generative AI today.

Although possibly apocryphal, French painter Paul Delaroche is said to have declared 'from today, painting is dead' upon seeing his first daguerreotype – one of the earliest publicly available photographs formed on a polished silver surface.[41] He expressed the widespread concern that photography would come to replace painting and undermine the teaching of traditional artistic skills. Other artists were critical of photography and didn't see it as a true form of art. In the 'Salon of 1859', French poet and critic Charles Baudelaire considered that photography's attempt to become an exact reproduction of nature came at the expense of an artist's imagination and creativity. He considered photography 'the refuge of every would-be painter, every painter too ill-endowed or too lazy to complete his studies'.[42]

An important shift gradually occurred as debates began to move from outright rejection and fears of obsolescence towards how photography could be seen as offering its own unique form of artistic representation. Pioneers such as Julia Margaret Cameron began experimenting with photography and attempted to elevate it to a 'high art' by adding details of dramatic lighting and soft focus on her subjects.[43] Cameron introduced a level of artistry into her photography and even included imperfections such as fingerprints, swirls and smudges as part of her art.

At the same time, photography began to influence painting and provided impetus for painters to reimagine their practice.[44] With photography dominating the realistic depiction of nature, painters began to move away from visual realism and experiment more with representations of light and colour, creating new artistic movements including Impressionism, Symbolism and Tonalism. Other artists incorporated aspects of photography into their painting, such as French modernist painter Édouard Manet, who cropped some of his images in a similar style to photography and incorporated more realist elements into his painting. Painters also used photography as a tool to assist them in their own craft: photographs could be a useful way of capturing an image to work on a painting in a different location or without subjects needing to remain still.

Rather than a simplistic view of one creative technology superseding another, the picture is often more complex and nuanced. Photography was eventually seen as its own unique form of art with its particular style and skills. Other art forms did not fade away, but were instead invigorated by the invention of photography and moved in radically new directions as a result. Some professions such as portraiture were largely made obsolete, but in their place photography was eventually made available to a broader range of people who would soon be able to take pictures with simple equipment whenever they wanted.

Similarly, AI can be used by artists to enhance their practice, offer them new ideas and help actualise their vision in ways that were not previously possible. Contemporary artists such as Anna Ridler understand AI as one possible tool among others that they can use in their practice.[45] For her, these machines open new possibilities that would not have been available without them. While they cannot act autonomously, under her direction they can produce something completely new in the process that is a combination of human and machine. Helena Sarin is another visual artist who uses machine learning techniques in her art, more specifically GANs that allow her to combine art and software to create new, arresting images.[46] The originating power of creativity still lies on the human side, but the final product is a collaboration between the two. AI could also be usefully incorporated into artists' practice in a more minor role, as a tool that could spark ideas and help organise work. Sudowrite launched an AI tool for writers called Story Engine which helps writers structure narratives, come up with ideas and write chapters. The tool has been widely criticised by some writers as degrading their craft, but from another perspective, it could be seen as a tool to help unlock writers' creativity and escape writer's block.

One argument that is often made in favour of generative AI is that it opens up certain forms of artistic creation to a broader audience beyond the naturally gifted and talented. Of course, amateur art has always been possible for people, but these tools increase the possibility for almost anyone to create unique images and videos that would have been unimaginable even a few years ago. Ordinary people now use AI to produce more than 34 million new images every day.[47] Most of these users are not professional artists, but with AI they can produce stunning images in specific styles and genres. As *Wired* magazine put it, 'artificial intelligence has become an engine of wow'.[48]

The real risk of the AI revolution in art might not be of human art becoming obsolete, but of the technology being abused by powerful interests to further exploit artists and make more money for corporations. Big production studios are keen to automate as many processes as possible and rely on human creatives only when necessary and with minimal cost. How technology is used in practice always depends on complex social and economic factors. The problem with generative AI is that it is seen by many as a shortcut that can do things on the cheap and as a way to avoid paying artists their due.

Laura worried about the kind of world we would create if we relied too heavily on AI art. For thousands of years, humans have created art that has been incredibly valuable to them and the development of civilisation. We create culture, expand what it means to be human and leave something for the next generation to ponder how to live in an increasingly complex world. If AI takes over, all the important reference points will simply be human works of the past, with little creativity or invention of the new. Will we end up living in a fundamentally decaying society in which our ethical and aesthetic sensibilities are trained off copies of copies, with the originals receding further into a forgotten past?

Art has an intangible element that is difficult to define. It contains something of the human spirit within it that is lost in algorithmic reproductions. AI cannot reflect on the feeling of a perfect morning gazing at your soulmate sleeping next to you or the terror of fighting in the trenches of modern wars. We read *Othello* to reflect on our own jealous tendencies and listen to Chopin's *Marche Funèbre* to contemplate our mortality. Without an emotional connection, generative AI can feel empty and devoid of meaning. But more fundamentally, generative AI in the hands of powerful corporations will likely become a tool for the rich to keep lining their pockets at the expense of artists across the globe.

5

The Operator

Alex's alarm goes off at 6 a.m. His eyes still closed, he reaches out for his phone and feels for the power button to silence it. Rolling over, he glances at his girlfriend next to him, hoping that her sleep hasn't been too disturbed. Every night he leaves his work clothes in the bathroom of their two-bed rented house so he can get up before sunrise without waking her. Each morning, the goal is to shower, eat, and be out of the house in thirty minutes. He pulls out of his drive, turning left and heading out through the newbuild estate he lives on, past rows of small terraced houses with fake plastic lawns. The inside of the windscreen is covered in condensation and his ageing car is slow to heat up. He uses a cloth to wipe the glass. He's on his way to work in a warehouse at the forefront of the logistics revolution: a cavernous hall full of whizzing conveyor belts and tonne after tonne of commodities, the dance of which is choreographed by one of the most complex AI systems in the world.

He drives to work through Coventry, a city in England's West Midlands that used to be a major hub of Britain's car manufacturing industry. By the 1960s, the intense bombing of the city during the Second World War seemed like a distant memory. Coventry had become known as 'the British Detroit', such was its industrial prowess. In 1971, the West Midlands was the second richest region in the country, and responsible for 75 per cent of total UK car production.[1]

By 1974, a colossal 52 per cent of the local population worked in car manufacturing, amounting to 115,000 people.[2] But the optimism didn't last.

The deindustrialisation of the late 1970s hollowed out Coventry in a way that the bombs never managed to. In 1981, local band The Specials released a song about their hometown that spent weeks at number one on the singles chart. The title? 'Ghost Town'. By 1982, 53,000 people had lost their jobs in the automotive industry.[3] What remained of the industry limped on, until the closure of the Browns Lane Jaguar plant in 2005 marked the end of 110 years of continuous car manufacture in the city.

It's a story Alex knows well, because it's his story too. Many of his mates at college had done apprenticeships at Jaguar, working as 'track rats' on the assembly line (aka 'the track'). They were either transferred or laid off when the factory closed. Some manufacturing stayed on, though. Bits of the old Browns Lane factory were rented out to subcontractors, and that's where Alex found work as a machine operative for a company producing car interiors that were still sold on to Jaguar. But it wasn't to last: in 2017, Alex had been one of 500 to lose their jobs when the company shut down for good.

In 2018, a new Amazon warehouse opened on the site of the old factory. Lots of ex-Jaguar employees ended up working back on the same site. Alex joined them – he didn't have many other options. It's a worse job than being a machine operative. He works longer hours for less pay and he feels the managers treat him like a child, but at least Amazon isn't about to go out of business. Amazon warehouses have occupied the post-industrial spaces and hired the post-industrial people that car manufacturing left behind.

Amazon gives all its logistical facilities code names based on the nearest airport, and this one is known as BHX4. It is somewhat unusual. Rather than just being a fulfilment centre, it supplies other warehouses within the Amazon system. Internally, it's known as an 'Inbound Cross Dock'. There are only three of them in Europe, and they occupy a strategic node within Amazon's distribution network. BHX4 is located an hour south of East Midlands Airport, the UK's second biggest cargo airport after Heathrow. It's in the UK's 'golden triangle', where 90 per cent of the population are accessible by road within four hours. This part of the Midlands, once a centre of the car industry, is today the country's logistical heart, home to a network of massive distribution centres. Driving along any of the major roads in the triangle takes you past one cluster of giant warehouses after another.

Amazon hides the details of its logistical network from scrutiny by the public and its competitors. The precise way in which this complex network of facilities and technologies combines is a mystery to most of its consumers, who only see the packages that arrive at ridiculous speed. But it isn't hidden from Alex. The operation of the network is his bread and butter. He pulls on his hi-vis vest and lanyard before walking across the car park towards the entrance, past the temporary CCTV installations that have been in place ever since union organisers started handing out leaflets on the road outside.

Alex arrives at 7.10 a.m. He has some time to chat with coworkers from his team in the canteen before the morning shift starts. He joins a table and they swap stories. There are something like 600 workers in the warehouse on any given day, all broken down into functional groups. The clock ticks down to 7.30 a.m., and the morning briefing that starts their shift.

The briefing always includes information about safety and quality. It must be boring to give the briefings, Alex thinks, because it's sure as hell boring to listen to them. Today's one has a bit of variety, though: a special section talking about the union. Apparently, they were leafleting the night shift as they arrived at 6 p.m. yesterday. Their team leader holds up one of the new posters that is being put up around the warehouse. It reads: 'Instead of joining a union, come and talk to us for free!' He says that Amazon values its employees, and that if anyone ever experiences a problem they can always use internal channels to get it sorted out. Management really cares about what they think, he promises. Alex exchanges a smirk with a friend. Sure they do.

Once the union discussion is over, the manager moves on to talking about volume – meaning the number of 'units' that need to move through the warehouse today. They want to do more than 1.4 million.

A woman, aged about forty-five, curses in Filipino. She turns to her friend and gestures at the ceiling. A few others shake their heads. The manager pretends he hasn't noticed. It's a higher target than usual because the last few weeks have seen supply interrupted by dockers' strikes at Felixstowe, the largest container port in the UK. About half of all containers entering the UK come through there, so the knock-on effects were significant and a backlog built up that affected the whole distribution network. Now they needed to pick up the pace to refill all the inventory that had emptied out over the past week. It will be a hard shift. Alex takes a deep breath, walks through airport-style security, and through the giant hall to his station.

To understand how the warehouse fits together, you need to understand the path of one of the 1.4 million units that will pass through it. Say this unit is a kitchen appliance, like a kettle. It arrives on the back of the lorry, packaged up with hundreds of other kettles on a pallet. Workers guide the lorry into a loading bay and a gang of three immediately gets to work taking off external packaging and loading items onto the conveyor to go into the warehouse. The goods they unload go on a conveyor down to a 'universal receiver'. This is where Alex gets involved. His task is to check a selection of the items and confirm that the AI system that rules the warehouse is receiving what it thinks it is. The system either tells him to just let them roll on down the line towards outbound, or sort them into yellow boxes or 'totes' that are arranged behind him. When those get full, the totes are automatically rolled away on a conveyor to outbound. There's no need to talk to anyone when you're on universal receive; you work alone with only the system to guide you. You just stand there and repeat that process all day long.

The totes and packages get labelled with their destination and sent through to be dispatched (sometimes by big industrial robots supervised by humans). On the outbound side of the warehouse they combine units for the same destination together into pallets, then wrap and label them again. Lifting the totes is hard work: on a shift like today you have to be doing a box every fifteen seconds or so. There are safe lifting protocols that restrict workers from raising totes above head height, but if you follow them it would be impossible to work fast enough. This is where a lot of the injuries in the warehouse happen. Another worker loads up the outbound lorry, ready for the goods to go out for delivery to fulfilment centres and for customer orders to come in. From there, they'll get sent to sortation and delivery centres located closer to big urban markets, where Amazon Flex platform drivers pick them up for final delivery.

As the day rolls on, Alex dreams of sitting down. They've never been given chairs on universal receive – or anywhere in the warehouse (apart from the managers' offices, of course). The idea is, apparently, that sitting encourages them to slow down. Everything has to move fast – because the system says so. Every role that involves scanning items is monitored closely by managers who are always checking the rate of units processed per hour. When Alex started working at BHX4, the measurement upset him. Early on, there was a day where he was really demotivated. The complete meaninglessness of the job had got him down, and he was working at about half his usual speed. At about 10 a.m., his team lead came over to him and asked what the problem was. He didn't have an answer, and the team lead stood behind him watching him work for something like five minutes, until he was confident Alex had sped up. It was humiliating to be so disempowered.

Before long, he realised that the rate was the main thing determining which of the agency workers got a full-time job and which ones didn't. He needed the job badly, so he worked like mad to get converted from the six-month agency contract to direct, permanent employment by Amazon. The rate stopped upsetting him as much after that because he learned how to comply: just work at an even, calm pace and let your mind drift off to something else. It was important not to clock 'idle time', meaning time you are meant to be working but your scanner isn't recording interactions with any items. You can only take the allotted breaks (two breaks of thirty minutes – one paid, one unpaid) and not a minute more. Alex's team lead walks round the floor at about 3 p.m., chatting to everyone individually about their rate for the day and if they had any idle time. He barely stops at Alex's station, just says 'great work' and carries on down the line.

Flunking the rate or logging idle time could mean you get put on an 'adapt', a performance management plan that you need to meet, otherwise you'll get fired. That used to happen when you were in the bottom 25 per cent for the rate five times in six weeks. Everyone wants to avoid that. The challenge is, however, that nobody really understands the rate anymore. They always used to announce the rate and update people on where they were at. They were told the exact number. In universal receive it was usually something like 300 units an hour to get 100 per cent. Each worker would have slightly different rates, so it wasn't always that simple, but 300 units was a good rule of thumb. A few months ago, however, they stopped handing round the rate paper with everyone's numbers on. Now they only tell the people who are a long way off the rate, in the bottom 25 per cent, to speed up. The problem with a relative target, though, is that even if they all sped up, there would just be someone else in the bottom 25 per cent. You can never all be safe. The slowest will always be missing their rate. Even if they did well over 300 units, they could never be certain. That lack of information means that they never know when they're safe, never know if they can relax for the last hour of the shift.

Alex has been in the same department for five years now. He wouldn't say he was depressed. It was more like a state of total demoralisation. Every time he walked into the building, he thought he needed to get another job. When he's working all day in an isolated role and not interacting with anyone, he gets questions buzzing round his head. The biggest one, always, is, 'What are you doing with your life?' He tries not to think about it, because when he does, the answer scares him.

Finally, the shift ends. He leaves his station and walks, very slowly, back into the canteen. There are jokes again, quieter this time. Alex used to ride a bike to and from the factory when he was assembling car interiors. There's no chance he could ride home now. His partner has a new joke; she's always saying, 'Your favourite word is tired.' He feels like he's become a different person since he started working here. More irritable, more miserable, harder to be around.

Meet SCOT, Amazon's Extraction Machine

When we imagine Amazon, we often think of Jeff Bezos, its former president and CEO, and how the company enabled him to amass a $170 billion fortune. Analysts often try to understand the success of a firm via the supposed genius of its founder. But to make sense of Amazon's retail operation, we need to start from an altogether more inhuman point: from a tangle of cables, servers, labour and data. Amazon's logistical network is built on the back of a collection of computer programs run by a team in Amazon called SCOT (Supply Chain Optimization Technologies). The team administers a system that is on the verge of becoming the assembly line of the twenty-first century.

'If the Amazon Store were a body, think of SCOT as its nervous system,' suggests Amazon's literature, describing a system which has 'end-to-end responsibility for orchestrating Amazon Store's supply chain' that is 'one of the largest and most sophisticated automated decision-making systems in the world'.[4] Amazon's supply chain network is made up of a bundle of different subsystems, each of which uses different computing techniques to solve specific problems. Overall, the system makes decisions and shapes outcomes across Amazon's whole supply chain, from the firm's interactions with manufacturers to the last mile of delivery. Just like the nervous system of the body, this system is 'quietly acting in the background to automatically optimize critical functions and flows'.[5] Its scale is enormous. It directs the organisational resources of a firm with an annual revenue of over $500 billion and a market capitalisation of $1.5 trillion. It may well be one of the most consequential deployments of AI anywhere in the world.

This supply chain management software has four major functions: forecasting demand; calculating inputs; planning fulfilment; and managing the fulfilment network. Each of these functions matches a different step in Amazon's logistical chain, from procurement to delivery, and has major implications for workers' lives. It centralises knowledge and decision-making in the hands of automated systems and high-level managers, disempowering workers in addition to deskilling and intensifying their labour in a manner which mirrors previous industrial technologies of control.

In terms of demand forecasting, the system runs one of the biggest simulation platforms in the world to predict peaks and troughs in the relentless flood of consumer clicks that generate sales.[6] Amazon's engineers have built and trained neural networks capable of forecasting demand across the millions of products the firm sells. Accuracy matters, because good predictions mean lower costs and bigger profits. Once it has this prediction of demand, the system then orders goods from producers and distributes warehouse space to vendors using Amazon's fulfilment services.[7] These ordering decisions determine the shipments that arrive at BHX4 every day. The system also identifies how the whole network should be resourced by equipping and staffing the web of trucks, warehouses and delivery centres that constantly hustles to meet customer demand. It can even make suggestions about where future warehouses should be located. By calculating the labour and material inputs needed to meet the forecast, it sets the stage for the system to generate a fulfilment plan.

The moment a customer clicks 'buy now' on Amazon's platform, the system uses real-time data and optimisation techniques to decide where to ship from and how to consolidate multiple orders in the most efficient way.[8] It dictates a dynamically optimised fulfilment plan that directs the whole network. It does this by using billions of data points, from page views and order numbers to barcode scans on warehouse floors, to determine how goods should be transported at maximum speed and minimum cost.[9] It manages the implementation of the fulfilment plan using a live feed of workplace data. Barcode scanning generates constant updates on where units are within the system and what processes they have undergone. The volume of scans per hour and per worker also gives the system detailed information on work intensity. This is used to generate a proprietary productivity metric. In 2019, Amazon internal documents stated that the automatic systems that track productivity automatically fire employees who consistently fail to hit targets or log too much idle time.[10] This individual productivity management overlaps with job rotation. In 2020,

Amazon announced plans to employ its machine learning algorithms to set workers' schedules according to muscle use and rotate employees among jobs that used different muscle groups to keep them more productive.[11]

There are other sources of warehouse data, too, like the video feeds taken from the interlocking jigsaws of CCTV cameras that watch every square metre of the warehouse. The fulfilment centres that come after BHX4 in the logistical chain are often built around 'cages' in the middle of the facility. There, small robots shuffle mobile shelving units around between the pick and pack stations that line the edge of the cage. Workers hate these stations. You're totally isolated from your colleagues for ten hours, just repeating the same few steps. It's a particularly brutal role when you're on the night shift. A lot of people end up with anxiety or depression. Each station is monitored by three cameras, which use a computer-vision AI called 'Nike' to maintain a record of where stowers have placed an item on the shelving unit so as to help pickers find it quickly later on. The system interfaces with human workers by projecting coloured lights on the shelving unit indicating where items are or where they should go. Ninety-five per cent of the time the monitoring system extracts the data it needs to work without human intervention, but in the other 5 per cent of cases the footage is beamed to low-paid annotators in India and Costa Rica, who work relentlessly for low pay to make up for the system's shortcomings and make sure it doesn't lose track of any items.[12]

The Bureau of Investigative Journalism has also found evidence of annotators being asked to conduct surveillance of workers to check Amazon COVID protocols were being followed as part of a programme called 'Proxemics'.[13] Amazon has also attempted to procure an anti-union surveillance system called the 'geoSPatial Operating Console' (SPOC) that tracks organising efforts, although Amazon continues to claim that it respects employees' rights to join unions.[14] Out on the roads, Amazon employs the 'Netradyne Driveri' driver-monitoring system across large parts of its delivery fleet.[15] The AI-enabled software uses in-cab and on-vehicle cameras to track drivers and give them instructions on how to drive and work. The exact dimensions of the company's sprawling surveillance programme are hard to determine. All the workers in BHX4 know is that they're being watched, tracked and monitored every second of the working day. The process of management is not completely automatic. Amazon workers are directly managed by team leaders and general managers in their workplaces. But the inputs these managers receive and the plans they work to complete have been determined at a higher level, in large part by the AI-enabled system.

Given this range of functions, it's not surprising that Amazon wants to present its AI system as a logistical brain. The metaphor puts the decisions of the system beyond scrutiny and inspires confidence in its technological sophistication. The illusion of the logistical brain helps achieve the support of managers, investors and regulators, and the compliance of workers. The metaphor continues a long tradition of understanding AI as a human brain. The birth of the field of 'artificial intelligence' is usually dated to a 1956 summer seminar at Dartford College, New Hampshire.[16] The name itself implies that AI would work by artificially recreating the processes of human thought through computing. This very old metaphor of calculation-as-thought both mystifies and naturalises how AI systems work. It encourages us to think of the decisions taken by these systems as beyond investigation or challenge because they represent a form of hyper-logical, quasiinfallible thought. But as we have set out in this book, it's much more accurate to think of these AI systems as an extraction machine. Just like every other system we have examined, Amazon's system relies on human labour and physical infrastructure. The intelligence of the system is extracted from the work activity of millions of warehouse and delivery workers, whose daily work creates the data points that power AI. It is their scanning of barcodes that generates some of the most fundamental information that the system relies upon. Just like a stoker used to feed coal into a steam engine, these workers feed data into the Amazon extraction machine.

There is a striking historical parallel between this system and the introduction of assembly line work in the twentieth century. The assembly line changed manufacturing forever. Before it, cars were manufactured by highly skilled assemblers who would crowd around a single vehicle and cooperate on completing each step in turn. They set the pace of work themselves. But the managers and engineers of the Ford Motor Company in Detroit had a different idea. Between 1909 and 1914 they experimented with a new approach to automotive manufacture. Henry Ford expressed its basic principle as: 'the man must have every second necessary, but not a single unnecessary second'.[17]

Over five years, the company introduced a number of major innovations: a new factory layout, a stricter division of labour, standardised and interchangeable parts, and single function machines organised in the order dictated by the production process. In 1913–14, these progressive improvements were crowned with the final touch: a moving line that moved the cars being assembled from station to station in a continuous flow. This was the assembly line. It was, in essence, an endless conveyor that brought cars past fixed stations where workers carried out simplified operations. Workers now only needed to specialise in one small task they would repeat over and over again, all day long. The pace of work was set by the speed of the line. Ford used the technology to achieve a new organisation of labour that deskilled and intensified the labour process. The time taken to assemble a Model T went from twelve hours to just ninety-three minutes, a 775% improvement in productivity.[18] The assembly line was the starting gun for both the creation of a vast manufacturing empire and the first car-centric society.

In 1914, Ford introduced a $5 day for workers who passed his strict company surveillance of their personal 'morality'.[19] The goal of the policy was to retain workers for longer (turnover had hit 380% the year before) and create a bigger consumer market for the Model T. As the car's price plunged, the company wanted to be able to sell cars to the working class market for the first time. But even this 100% wage increase was only a small step towards sharing the colossal profits enabled by the assembly line. Ford continued to plough these profits back into expanding production, and left workers to deal with the negative consequences of losing control over their own work processes. This contradictory dynamic has emerged over and over again throughout the history of modern industry. Rather than advances in productivity liberating us from work, the benefits of such innovation tend to go primarily to investors and owners. As Charlie Chaplin put it in the climax of his film *The Great Dictator*, 'Machinery that gives abundance has left us in want.' This dynamic creates a situation in which technological development in the workplace presents itself as an ongoing bosses' offensive. Workers, quite rightly, understand that the path of technological development prioritises the interests of the owners of capital, not the people who do the work. Opposition to managerial domination manifests itself as an opposition to technological change, because in this economic system one directly accompanies the other.[20] We could imagine more emancipatory ways of relating to productivity drives, but this would require an entirely different form of society.[21]

The workers who fled Ford's Highland Park factory in their droves in 1913 were fleeing for jobs in other plants where the assembly line was not yet in operation, and where they retained some degree of control over the work process. But over time, the assembly line diffused to become the fundamental technology structuring car manufacturing in the US and around the globe. There was nowhere to flee. Millions of workers were swallowed up. The car worker had gone from a skilled mechanic to an automaton, all because of the reorganisation of work by technology. The assembly line was not a unique workplace technology, but part of a much broader process that sought to centralise knowledge and decision-making in the hands of managers.[22]

This centralised knowledge is now located among a few people in Amazon's offices, blessed with access to the extraction machine's control interfaces. But for workers on the warehouse floor or in delivery vans, the AI-driven automatic system can feel like an alien force. It articulates them like a puppet, shaping their actions to an incredible degree. The workers in the network are reduced to the system's appendages – useful insofar as they are cheaper than industrial robotics and better at fine motor control and advanced problem-solving, but a problem insofar as they can deviate from the system's plan.

If human workers fail to be disciplined enough or cannot maintain the necessary pace, then they are cycled out for a new worker with almost no interruption. The job is easy to train on, and the dismissed worker will not have accumulated knowledge of any value. As a result, Amazon's turnover numbers have at times been almost as high as Ford's. But this creates problems of its own: a leaked 2021 report from the US showed that Amazon was on the brink of a hiring crisis in Arizona and California because of the speed at which it had burned through the states' workforces.[23] Amazon responded to this leak, stating the documents were 'most certainly early drafts that weren't appropriately refined or vetted, let alone finalized', but did so without clarifying specific figures.[24]

Whereas Browns Lane's old Jaguar car factory once saw workers organised in a series of sequential stations along an assembly line completing specialised tasks as part of a manufacturing process, now the same warehouse has workers spread out in a complex chain of interlinked processes and conveyor belts all coordinated by Amazon's system to meet a fulfilment plan. The data produced by their work becomes a fundamental input into the extraction machine that allows it to create the plan and manage a network that is approaching unparalleled levels of logistical complexity.

For workers such as Alex, the system determines the parameters for every moment of his working life. But just like the other workers in Amazon's global fulfilment network, he knows almost nothing about how it operates, and his desires and needs are not prioritised in its logic. Many believe the system is optimised for speed and cost, not for wellbeing.[25] His needs come second to those of Amazon and its customers.

Your Future Boss Could Be a Machine

These AI systems do not just shape the lives of the people who work at Amazon. They are an advanced example of a possible future of work that threatens billions of people. In previous chapters, we explored the supply chain of how AI systems are produced and the millions of workers involved in this often hidden process. Here, we turn to how AI is being deployed in workplaces in ways that are likely to have lasting consequences for an even greater number of workers. The primary value of the extraction machine for bosses in many low-productivity industries is the way in which it can squeeze more out of workers.

Some workers have always been subject to intense managerial control. The assembly line was, as we have seen, a way of increasing the pressure on workers in car manufacturing. The technology soon generalised to a whole range of manufacturing workplaces. More recently, digital surveillance was pioneered in the call centre, where for decades workers have had all their calls recorded for 'training and monitoring purposes'. AI management technology is a new iteration of this old dynamic. Rather than an entirely new development, these systems are extending existing patterns of managerial domination into new segments of the workforce. The proliferation of the extraction machine across industries, many of which are traditionally white-collar, has the potential to create profound consequences.

Workplace surveillance practices have exploded since 2020, with many companies now collecting information on almost every aspect of workers' jobs – sometimes without employees even knowing it is taking place.[26] Machine learning algorithms allow companies to analyse this data and predict future patterns of behaviour. These technologies surged in popularity during the pandemic as managers everywhere came down with a case of 'productivity paranoia', with 85 per cent of managers worried their stay-at-home workers were not being productive enough.[27] During the pandemic, many employers forced their workers to install various 'bossware' tech on their home computers and began monitoring their workday through a variety of apps and trackers. The number of large employers using such tools has more than doubled since the beginning of the pandemic.[28] In the US, the *New York Times* found that 80 per cent of the ten largest private employers track the productivity of individual workers.[29] Workplace surveillance tech is often pitched as a means of maintaining a safer and more harmonious work environment for workers. But the creep of surveillance and productivity tools into every aspect of workers' lives risks creating a stifling culture of overbearing managerial control.

In some offices, employee ID cards now track when workers come into the office and for how long they stay.[30] Other workplaces mandate that remote workers must keep their cameras and microphones on at all times during their work day.[31] Others still use surveillance software that can record almost anything a person does on their computer, from how much time they spend on particular screens to their typing rate and even monitoring them secretly through their webcam. In a report by US non-profit research institute Data & Society, researchers found that while many workers were aware they were being monitored, they often did not realise the extent of the surveillance or what was being done with the information collected.[32] In one example from the report, Walmart employees were asked to install an app on their phones for inventory checking, but the app had constant access to the phone's location and was continually sharing information with the company. There is a danger that as bosses get a foothold into their employees' private lives, this software will enable increasingly stringent forms of monitoring and control. History has shown that once increased surveillance measures are introduced to deal with a crisis they are rarely reversed – just think of the Patriot Act after 9/11 in the US, increased CCTV monitoring after the 2005 London bombings, or France's state of emergency measures after the 2015 terrorist attacks.

This technology is not confined to specific sectors of the economy or a few lone companies. It has proliferated across industries, as dozens of smaller tech companies have created apps and platforms to monitor workers' activity. A report by civil society organisation coworker.org found that over 550 different tech products had been created between 2018 and 2021 to digitise and monitor all aspects of workers' lives, from recruitment to disciplinary proceedings.[33] Previously, those who experienced the cutting-edge of this technology were gig workers, particularly in the transport, food delivery and care sectors, who had every second of their jobs quantified through practices of algorithmic management. A lot of these early experiments focused on the coordination of workers to complete tasks, rather than the supervision of their performance. But there were experiments with automatic supervision. Uber, for example, has been known to use access to drivers' smartphones to monitor them, such as checking their braking and acceleration habits as indicators of driving safety.[34] What we are now witnessing is these tools disseminating across the workforce as other professions are atomised, deskilled and monitored through productivity software. These include tools such as Hubstaff, Time Doctor and FlexiSPY, with 80 per cent of this software incentivised for long-term use on employees.[35] They often combine a growing mountain of data on workers with the predictive capacities of AI to increase managerial control over the workforce.

Most workers hate this form of surveillance. A survey of 750 tech workers found that half would rather quit their jobs than be subjected to constant surveillance from their bosses.[36] Some workers have found low-tech forms of resistance, such as mouse jigglers that mimic human movement to fool productivity monitors or software that disrupts spyware on their computer. But not all of this surveillance technology can be blocked or evaded. Use of some of the most invasive of these tools has increased dramatically over the past three years. Keystroke recording, which records everything you have ever typed, including passwords, has increased 40 per cent; stealth mode, which allows bosses to monitor employees without them knowing, has increased 38 per cent; and over a third of tools can now track employees' exact GPS location.[37]

The risk is that bosses' desire for increased productivity at work bleeds over into workers' private lives, with unprecedented invasions of their personal privacy and bodily autonomy. In addition to standard screen monitoring software and productivity tools, AI has enabled the creation of a new generation of *Black Mirror*-esque tech such as 'mood and sentiment analysis', facial recognition and voice-monitoring software. Employee monitoring software Teramind is one of the worst offenders, with the ability for bosses to watch or listen into their employees' conversations both in the workplace and even when they are at home through controlling cameras and microphones on workers' devices.[38] More specifically, once a rule violation has occurred, the company can record five minutes before and after the violation, including every keystroke and with full video. In addition, workplace wearables record employee health stats and enable the constant monitoring of individual workers to ensure they are focused and not too stressed.[39] Predictive AI tools use data collected on workers to rank them and give different individuals 'risk scores'.[40] One example of this is HR analytics tool Perceptyx, which produces a 'vulnerability score' of how likely a worker might be to unionise or leave the company. Perceptyx boasts a range of tools including 'Sense', 'an always-on employee listening product'.[41] This software generates an 'Engagement Index' score of individual employees based on an ana-lysis of their voice recordings and promises 'to reduce unwanted turnover through predictive modeling'. AI-powered text analysis software can also monitor employee emails and Slack messages to assess workplace culture and employees' sense of belonging at the company.[42] It is a short leap to a scenario where every minor facial expression or change in the tone of one's voice gets recorded and fed directly into an AI-driven scoring system, which classifies workers based on their mood and levels of employee satisfaction.

AI has also been widely adopted by HR teams for the purpose of recruitment and hiring. AI systems can be employed at different stages of the recruitment process, from sifting through CVs to creating shortlists, conducting interviews, assessing personality tests, and judging candidates. Between 35 and 45 per cent of American companies now use AI as part of recruitment, which has led to these companies cutting their recruitment costs by 40 per cent in 2022.[43] AI tools save HR time and money by automatically scanning applications for keywords and phrases. They also promise to remove the element of human bias by not focusing on a candidate's gender, race or age.[44] However, many of these claims are illusory, as the software itself has been demonstrated to show significant biases and further dehumanises the experience for job candidates.

Amazon offers another important cautionary tale about the dangers of AI recruitment. In 2014, Amazon engineers in Scotland created a program that would automatically rate job candidates and provide human recruiters with a list of the top applicants.[45] In 2015, the team realised this software was biased against female candidates, particularly for technical roles such as software developers and engineers. This bias was introduced through a gender imbalance towards men in the training data of previously successful candidates' applications. Learning from these past examples, the algorithm deduced that male candidates were preferable and downgraded applications that included the word 'women's' (such as in 'women's football team') and candidates who had graduated from women's colleges. When the algorithm searched for particular words and phrases in CVs, it also preferred action-oriented language such as 'executed' and 'captured', which is more likely to be written by male candidates. After unsuccessfully attempting to fix these biases, Amazon scrapped the project in 2018.[46]

Researchers at the University of Cambridge have studied the potential for AI to eradicate forms of race and gender bias in the recruitment process and found significant limitations in what software could achieve.[47] The researchers convincingly argue that such tools reveal a 'prevailing misunderstanding of what race and gender are' by viewing them as potentially removable attributes of an individual rather than broader systems of power that shape how organisations and society operate. As the Amazon example demonstrates, racialised and gendered systems influence the type of language individuals use, how they behave and how they are perceived by others, even when purportedly neutral algorithms are performing the assessment. As a result, an algorithmic system that is trained only to see personality traits, concrete skills and the prevalence of keywords in CVs can still produce discriminatory outcomes.

One metric used by many of the AI systems is an assessment of a candidate's personality using the 'Big 5' personality traits (extraversion, agreeableness, openness, conscientiousness, neuroticism) in an attempt to avoid any racialised or gendered categories. These AI companies claim personality scores provide a neutral assessment of a candidate's aptitude, which is more meritocratic than a recruiter's subjective opinion of their experience. Yet even such personality tests do not seem to produce unbiased results. Evidence has shown that what a candidate is wearing and which language they speak during the test can change some of the tools' perceptions of a candidate's personality.[48] These so-called personality assessments verge on a pseudoscience, and make outlandish claims to predict job performance by creating reliable 'behavioural profiles' on people based on the tone of their language or their on-screen appearance.

Companies are not only using software to sort through resumés. Some are inviting candidates to automated interviews in which they are requested to stare into their webcams and answer pre-recorded questions. Their responses are then analysed by AI-driven software to determine their suitability for the job. Modern Hire, for example, boasts an 'AI-powered automated interview platform' it calls 'Automated Interview Creator' that helps HR departments develop interviews and assess candidates.[49] AI tools are used primarily to transcribe the interviews and to evaluate and rank candidates based on the interview text. But previously, HireVue (which acquired Modern Hire), used facial recognition AI tools to analyse video interviews, resulting in a complaint filed to the Federal Trade Commission for engaging in unfair and deceptive practices.[50] This software claimed to analyse candidates' facial expressions, microgestures, tone of voice and online presence to assess personality traits and suitability. While HireVue stopped using these tools, it's difficult to say how prevalent they remain in the sector. Some US states are developing laws to regulate the use of AI for recruitment, such as the Illinois legislature's Artificial Intelligence Video Interview Act and New York's Automated Employment Decision Tool law.[51] The EU AI Act also classifies AI-enabled recruitment software as 'high risk', meaning companies will have to meet a long list of requirements to ensure it is safe and provide information on their system in a publicly accessible database.[52] But for the time being, much of the use of this technology remains up to the employer, leaving broad scope for potentially biased decision-making.

The deployment of AI technology in the workplace has the potential to advance new forms of surveillance and managerial domination. Layers of white-collar workers who used to experience more autonomy in how they organise their work are increasingly subject to practices that treat them just like Alex and the track rats that came before him. Work intensification is spreading from industrial sectors to retail, hospitality and the wider service sector. All of these control technologies have a dual purpose in the workplace. They both cut the costs of production (at the expense of job quality, workers' autonomy, and basic democratic rights) and empower bosses at the expense of workers.

Surveillance tech enables employers to crack down on worker resistance and also has the potential to impede worker organising and representation within companies. The ability to collectively organise is enshrined in most countries' laws, but exercising this right has become increasingly difficult as companies engage in sophisticated forms of union-busting. Bosses monitor worker conversations on workplace channels or spy on workers via cameras in addition to hiring specialised agencies who advise them on how to prevent their workers from forming trade unions. Amazon, for example, was accused by its employees of quietly monitoring email lists of those employees who were involved in activism around union activities.[53]

Workers resist the deployment of this kind of tech in lots of ways, from covert rule-breaking to leaving the job. Most of these forms of resistance are individual or shared with a very small group. Five or six coworkers might discuss how to scam the rate calculation in an Amazon warehouse and earn themselves some much-needed rest, but this kind of subterfuge usually happens in secret. We can be certain that tactics like this are already widespread at Amazon, even if we have no direct evidence for it – because such resistance is always taking place. It is the organic by-product of the clash of interests between workers and bosses in the workplace. But sometimes, when certain conditions are met, this informal and individual resistance scales up into something more durable and overt. That is exactly what has happened in Coventry at BHX4.

The Strikes

You'd be forgiven for assuming that systems like Amazon's fundamentally and irreparably shift the balance of power in favour of managers. The accumulation of data and knowledge made possible by digital surveillance and the complex management facilitated by AI seem like they could block any avenue for resistance. When researchers conclude that Amazon implements a system of 'techno-economic despotism' it's hard not to feel a pang of despair.[54] But in fact, recent history shows us time and again that the most intensive modern management techniques do not eliminate workers' ability to fight for their own interests. Instead, they force that resistance to take on new forms.[55] After all, that's what happened with the assembly line: at first, workers were undermined and disorientated, and trade unionists predicted that it would be impossible to organise against it.[56] But then, workers in the US developed the tactic of the 'sit down' strike, in which they occupied their factories for the duration of the action, and led one of the most incredible unionisation drives in the history of the global working class.[57] The future of work managed by AI is not just a future characterised by an increased potential for managerial domination. It is also one in which workers invent new forms of resistance.

Workers' struggles in the platform economy have offered a preview of the fights to come against AI management. Workers for companies like Deliveroo and Uber have launched guerilla campaigns to fight for higher wages and better conditions. They have employed a complex mixture of wildcat strikes, protests, unionisation and more covert resistance in a wave of action that has spread almost as fast as the platforms themselves.[58] Emphatically, the introduction of algorithmic management to food delivery or taxi work does not eliminate workers' ability to fight back against exploitation. These lessons are in line with what we have found during our research at Amazon. After a year of interviewing workers, visiting warehouses and standing on picket lines across the UK, we haven't found anyone who could be described as the passive victim of a system of total control. Instead, we have found a workforce that is continually engaged in forms of self-activity that are calculated to defend their interests. They take both covert and overt action to resist their bosses and their systems of control.[59] They find ways to take unauthorised breaks inside the warehouse; they quit the job in disgust; they build friendship and solidarity with their coworkers; they sabotage conveyor belts; and they go on strike. Far from eliminating the working class as a political actor, Amazon is building a new working class in the deindustrialised zones that surround the major cities of Europe and North America.[60]

The union had been outside BHX4 for years, on and off. They stood on the road leading into the warehouse trying to talk to people as they drove in. Hardly anyone ever joined them. Alex didn't stop for a chat because he was worried about getting hassled by managers driving in behind him. Once or twice, he took a leaflet handed to him through his window and gave it a scan, before sliding it into the pocket beside his seat. The leaflets collected there, one after another, waiting for the day that he would finally remember to chuck them out. He didn't give the union much more thought until the summer of 2022. The Russia–Ukraine war caused prices to start going up fast, and suddenly his wages felt like they covered a lot less. Every food shop was getting more expensive, but his pay packet stayed the same. He began to see his colleagues work more and more overtime. Some people were working two extra shifts a week. They went right up to the maximum of sixty hours a week to hold on to their standard of living. Others finished their shift, walked out to their cars, and logged in as Uber drivers. Week after week he saw more cars parked outside the warehouse with an 'Uber' licensing sticker on the driver's side door. Inside, everyone was getting more stressed. Disagreements that would have been resolved with a laugh six months ago turned into shouting matches.

Then, without any warning, it boiled over. In August 2022, the news said inflation was at 10 per cent. A pay rise was coming, and everyone expected it would be at least £2 an hour. In August, the announcement came: pay would go up, but only 50p an hour. The pandemic had seen Amazon's profits grow by 220 per cent, but apparently there was no money left to cover the cost of living for the people who made those profits possible. Protests broke out at two warehouses, and videos started circulating of what had happened at Tilbury in Essex. There, hundreds of workers had walked off the job together and gone to talk to the general manager. Their startled team leads told the crowd to meet in the canteen for a proper conversation, where a group of nervous managers attempted to calm things down. They claimed they knew nothing about the new pay rate, and they would find out more soon – but that everyone had to go back to work. If they didn't, they wouldn't get paid. The crowd refused, and an informal strike began. When the night shift started arriving, they joined in too.[61] The idea spread fast.

The next day there were strikes at five warehouses, from Tilbury to Bristol, and serious disruption at four others. At BHX4, the seed was planted during a 10.30 a.m. break. A group of workers watched TikTok videos of the walkouts. They decided they would start their own one at lunch. Alex heard the commotion from his station and joined in. Within minutes, 300 people had walked out and were refusing to work. They sat in the canteen, joking and sharing videos of the strikes in other warehouses.[62]

For the next three days, workers at BHX4 walked out over and over again. On the second day, the General Manager (GM) asked to speak to the leaders of the strike. The strikers refused: management could talk to all of them or none of them. After some argument, the GM agreed. The workers asked questions about how the 50p raise was decided and demanded a bigger increase. The GM promised to 'take it away and try and get an answer', before reiterating that nobody would get paid unless they went back to work. The strikers refused to go back anyway. The capacity of managers to give orders had completely broken down. And the same scene was repeating itself across the country. On the eighth and ninth days, workers at the mega-site in Swindon were out on strike. Amazon bought the warehouse for £200 million in 2020. At the time, it was the single most expensive logistics property ever bought in the UK. It cost at least as much again to fit it out according to Amazon's requirements. But now, all that capital lay idle, because without human labour, nothing worked. On the tenth day, another four warehouses were on strike, and then finally one more walked out on the eleventh. But the wave wasn't quite over yet. Over the next few days, it spread over thousands of miles. Workers at an Amazon warehouse in San Bernardino, Southern California walked out over wages and workplace temperatures. They weren't alone – in Turkey, 600 workers at Amazon's main Kocaeli warehouse launched a sit-down strike, and in Germany, workers launched an official strike over the rising cost of living.[63]

By the time the surge of action came to an end, there had been at least twenty-two actions at eleven workplaces in the UK alone. There was no precedent for action like this. There had been wildcat strikes at Amazon warehouses before in Poland, France and the USA,[64] but they had never spread so fast, or been so determined. Amazon is an anti-union employer: in country after country, from the UK to Poland, France, Spain, Italy, the Czech Republic and the US, the company has consistently resisted collective bargaining, and only conceded when forced to.[65] Amazon managers have successfully seen off union organising drives in city after city.[66] To do so, they follow a very effective playbook: in the UK, the first effort at unionisation collapsed in disarray in 2001.[67] Now, however, it seemed that the company was taken by surprise. The form the strikes took was probably one of the reasons why. The wildcat strikers had no legal protection against dismissal, but they also skipped the legal obligation to give their employer advance notice of their action. They didn't follow the UK's restrictive trade union laws, which meant that the strike was unofficial. They were relying on solidarity to protect them. And it worked. After the strikes ended, there were no mass layoffs. Instead, there was the start of a new drive to unionise Amazon warehouses in the UK.

The strikers' action made one thing very clear: Amazon workers are a powerful, combative force. After years of being represented as the passive victims of all-powerful managers, they finally had their say. The wave of strikes kickstarted a wider process. There had always been a group of union members at BHX4, but now that initial group started to grow. After the wildcat action lost momentum without winning a pay rise, people became convinced that the union could help them take the next step in the fight for better wages. The walkouts started to become mythologised as the start of something: the first moment they had really fought back. Alex wasn't so sure. His dad was a union man back in the day, but they'd been beaten. That whole era seemed like the distant past. Thatcher had won; the pits had closed and the factories had shut down. There was nothing left to fight for. He pinned his hopes on making his way up the promotion ladder, or maybe getting a new job if one of the big manufacturers came back to Coventry.

It was a conversation with one of the older lads he often talked to in the mornings that finally convinced him to join. He was about sixty, and looked like Alex's uncle. It was a Friday morning, and Alex had asked him if he had any plans that weekend. The bloke sat for a second quietly. 'No plans, no.' He shifted in his chair and stared down at the surface of his cup of tea. He explained that he used to work forty hours a week over four days, take a day to go fishing and spend the remaining two with his family. Now, though, he needed to work five or six days a week to make ends meet. 'The way things are, I can't afford to go fishing anymore.' That evening, Alex joined the union on his phone as he was walking back to his car.

The wildcat strikes proved one unavoidable fact: despite the complexity of Amazon's systems of control in the warehouse, workers were not totally disempowered. Strike after strike followed, and the union continued to grow. Alex went out on his first official strike, then his second, then before long it became routine. By the end of 2023, the workers at BHX4 had been out on strike for over thirty days in total. Their union, GMB, was starting to talk about using a statutory mechanism to force Amazon to start collective bargaining with them over wages.

A few days after the 1.4 million-units day that began this chapter, Alex is out on strike again. It is dark in the mornings now. The time is 6:30 a.m. and he is standing in the road, rain hammering down around him in the kind of fat droplets that soak you through in seconds. On the road outside the warehouse, baby-blue lorries with giant Amazon logos on the side have been slowed to a halt by the traffic. In between them sit the cars of people trying to break the strike and go into work. Alex walks up the queue of vehicles, noticing how the cops keep on running backwards and forwards in an effort to get things moving. Bob Marley plays from a distant speaker, and someone somewhere is chanting on a megaphone. There are about

400 workers spread out up and down the only access road into the business park. This is the union hardcore – the ones who turn out no matter what. He recognises most of them. He doesn't get on with everyone, but at least they're outside. A union staff member gave him some leaflets earlier, but they've turned to mush in the downpour. Looking to his right, he spots someone he recognises sitting in their car, and he gestures for them to roll down their window.

'Hi, mate. You know this is a picket line, right?'

6

The Investor

The dashboard on Tyler's screen is full of upward trajectories and green lights. Everything is, as far as he can tell, going well. The charts in front of him show a consistent rising trend in users, an expanding number of daily requests, and a series of 'on track' indicators next to development milestones. These are the vital signs of a company beginning a rapid period of growth. That's a good start – but it isn't enough. He needs this investment to really fly. His overall numbers have remained stubbornly below the team average for the last year. He's not the worst performer, but he's not the best either – and he can't stand it.

Tyler is a partner at a Silicon Valley venture capital (VC) firm. His job is to pick winners: to find and invest in tech startups that are on the verge of massively increasing in value. He grew up in the Inland Empire metropolitan area of Southern California, where his mom raised him alone without much support from her family. She had moved out of Los Angeles in search of cheaper housing, and found work as a veterinary nurse in San Bernardino. Tyler came of age in the lower half of the middle class. He'd never been exposed to the worst poverty, but his cousins back in LA had always been better off. He got in some scraps with the kids in his neighbourhood, and the experience had marked him out as a different kind of person than the rest of his family. He felt like an outsider at the barbecues they had by his uncle's pool. At high school, he ran cross country, and despite an obvious lack of talent, he did all right. It wasn't unusual to see him collapse as soon as he crossed the line. He'd lie on the floor, chest heaving as he tried to recover his breath, his face planted on the short grass of a local golf course.

Tyler took a strange route into the VC industry. A human rights lawyer by training, he ended up changing careers via a series of co-incidences. One of his college friends began working for a financial technology startup after graduation. The company raised tens of millions from venture firms, and before long the startup's young employees became close friends with some of their investors – so close they went to the annual Burning Man music festival and took a couple of tabs of acid together. Tyler was there too and joined in, and a few weeks after he found himself wondering if his entry-level job at a legal NGO was really the most effective way to improve the world. Maybe, instead, he should become an investor and help shape the future by directing funding towards startups with positive missions. It wasn't long before he handed in his notice.

The fund he works for controls $500 million of investment, and is focused entirely on the tech sector. Its goal is to produce a 300 per cent return on its investments over ten years through early stage investments in startups. Most of the firms they invest in will never turn a profit, but the few that do can provide huge returns, in the order of 1,000 per cent – enough, Tyler hopes, to triple the capital of the whole fund. His firm specialises in Series A funding, meaning they like to be involved in a startup's first big round of fundraising. It's the real-world version of the classic reality TV show *Dragon's Den* (*Shark Tank* in the US): startups that need more resources to grow use these fundraising rounds to get cash from investors looking to make big profits. If things go well, venture capitalists will realise these profits in one of three ways: either they cash out when the startup 'goes public' and sells shares on the open stock market, 'gets acquired' by a Big Tech firm, or, finally, they can sell their stake to another investor in a later funding round. Tyler's job is to use his networks to find interesting people working on interesting projects, assess which ones are winners and guide them to success.

He spins in his chair and goes to get a coffee from the kitchen. The routine is always comforting: grind the beans, prepare the machine, tamp down the puck, pull the shot, steam some milk – it takes him three minutes from start to finish, but for those three minutes he doesn't have to think about the numbers or check his emails. In half an hour, a company car will take him to his office in Palo Alto. Later, he's got a lunch meeting with a VC colleague to discuss the new startups emerging from Stanford University, the big incubators and seminars that bind the Silicon Valley community together.

He settles back in front of the dashboard. Its metrics are for a startup that is building a new AI health and safety product: a system that can use existing workplace CCTV video feeds to identify potential risk areas, near misses, and monitor compliance with health and safety rules. It's meant to increase the efficiency of health and safety reporting in workplaces like ports, transport hubs and warehouses. '5,333 employees are killed in US workplaces every year,' the website reads, 'but 100 per cent of these deaths are avoidable.'

One number on the dashboard stands out. It is a calculation of the gross margin, done on the basis of approximate numbers from the past quarter. That means it takes all the income of the startup, subtracts the costs of running the system, and generates a number that is tracked over time. Like all startups, it began deep in the red. It's slowly climbing towards the green, but Tyler is anxious about the gradient of that line. If this investment is really going to pay off, it needs to be a lot steeper. There are, he thinks, two basic options for how to make that happen.

First, increase the startup's income. Either sell more of the product, or sell the same amount at a higher price. The latter option won't fly: although it sounds like a niche product, this part of the market is already crowded. If they raise prices, someone else will just undercut them and take their market share. If they want more income, they'll have to find more users. To find more users, they either need to do a better job of winning the existing demand, or innovate with the product's functionality so they can sell to a wider range of customers. Tyler's been thinking about one option: right now, they have a system that watches for safety incidents and rule violations in the workplace. Surely it wouldn't be too difficult to train that same system to look for other kinds of incidents. Maybe it could highlight potential theft – then they could sell it to grocery stores, or even warehouses that were worried about 'shrinkage'.

Second, they could cut the product's cost. He pulls up a record of the company's expenditure for the last month and filters by amount. One item near the top of the list stands out: data annotation. As of that moment, they're paying hundreds of thousands of dollars to a firm based in Birmingham, Alabama. At their offices, workers are employed to watch hour after hour of footage and identify when specified safety rules are broken or accidents take place. They are given strict instructions to follow, and the data they produce helps increase the accuracy of the product. Tyler thinks for a moment. A data firm's number-one expense must be labour. So the best way to cut costs would be to find cheaper workers. He guesses that these annotators are paid $15 an hour. Well, there are billions of people in the world who will happily work for $15 a day.

Both of these options make Tyler uncomfortable. He sits, head in hands. Nobody else at the office would be worrying quite so much about this. They'd just pick the best option and go for it. He broods over the ethical implications of every decision he ever makes. Sometimes business has to be business. Fundamentally, he thinks, markets are systems for the rational allocation of resources. You can't fight them, or try to impose your own logic on them. If this company is going to thrive, then it needs to have the most rational utilisation of resources among a crowded field. After all, if the gross margins don't improve, then the company won't get acquired, he won't make his targets, and nobody will get the benefits of a safer workplace. He jots down a quick note in his phone: 'Data annotation – can we cut costs? Can we find a non-US annotation partner? East Africa?'

He'll write his thoughts up into a longer email in the car on the way to the office. They have team breathing exercises scheduled for mid-morning, followed by a sharing circle, then the lunch meeting.

The AI Gold Rush

So far, our approach to AI has been to go inside the workplaces where the technology is being developed or deployed to understand its implication from the workers' point of view. This chapter takes a different approach, and tells a different kind of story. This time, we take the elevator up to the boardroom, to look at those making the fundamental investment decisions that shape the future of AI. We will ask *what* business environment has been produced by the AI gold rush, *who* controls the process of AI development, *where* they are making decisions, *which* structural pressures they are making those decisions under, and *why* their vision of the future should concern us.

The AI hype train pulled into Palo Alto station at just the right time. Towards the end of 2022, Big Tech share prices had tanked, its major companies were shedding staff and none of its big ideas were sticking. At the peak of the pandemic, the Valley had been riding high as the world was forced indoors and online. A decade of technological adoption happened overnight as businesses and public services scrambled to adapt to lockdown laws and shift to a digital-first world. During a global downturn in the economy with plummeting share prices, tech stocks surged over 43 per cent during 2020.[1] Tech companies also seemed to momentarily enhance their battered reputation. Conversations about surveillance capitalism, privacy violations and tech monopoly power temporarily receded as companies rebranded themselves as 'digital first responders' to the pandemic – heroes that worked hand in hand with governments and essential workers to provide the digital tools needed to bounce back. But the good times didn't last.

As 2022 rolled on, the biggest companies all lost steam and were scrambling for ideas. Facing an economic downturn of higher interest rates and rising inflation, they were beginning to feel the weight of their skyrocketing operational costs. By November, Meta had announced it would be laying off roughly 11,000 people and would soon proclaim a 'year of efficiency'. Two months later, Alphabet reported job cuts of 12,000, while Amazon would be letting go 18,000.

Across 2022, a total of 93,000 jobs were lost across public and private tech companies in the United States, with layoffs increasing a whopping 649 per cent from the previous year.[2] Tech stocks were down 20 per cent for the year and the Bay Area looked like it was in a permanent doom loop.

Meanwhile, new ideas for the 'next big thing' were not paying off. Meta poured $13.7 billion dollars down the virtual drain of the metaverse in 2022 alone (as of October 2023, Meta's Reality Labs had lost a total of $46.5 billion),[3] as it discovered people did not want to live most of their lives in an 'embodied Internet' powered by virtual reality technologies. Following an investment the size of the Apollo space program, all Meta seemed to be left with was a low-res picture of a cartoonish Mark Zuckerberg hovering in disembodied space in front of the Eiffel Tower. Equally hit was the crypto market, which bottomed out with the collapse of crypto exchange FTX and some of its leading figures being hauled off in handcuffs, facing criminal charges of fraud and conspiracy to launder money. Looking back on the year in tech – web3, crypto, blockchain, bored apes and NFTs – so many ideas from this period feel like we were in some kind of futurist fever dream.

With rising prices squeezing consumers' ability to purchase services from platforms, and interest rate rises increasing the weight of corporate debt and disincentivising investors from taking big risks, new platform bets didn't look like a viable option. Yet all hope was not lost. Another story was unfolding about a new investment opportunity that might yet save the Valley. For the public, the turning point was the explosion of interest in generative AI in November 2022, in particular following the launch of ChatGPT and image generators that were already available, such as Midjourney and Stable Diffusion. A series of deals struck in early 2023, such as Microsoft's $10 billion investment in OpenAI, seemed like a turning point, but this was not Microsoft's first rodeo. It had invested $1 billion in OpenAI back in 2019, a lifetime ago in the world of tech, to become its preferred computing partner – so OpenAI could use Microsoft's Azure platform to develop the next generation of AI products. By the time it announced its mega deal in early 2023, it had already ponied up another

$1 billion in 2021. In fact, annual global corporate investment in artificial intelligence peaked in 2021, during the era of low interest rates, at a total of over $119 billion, which then dipped in 2022 to just a little over $64 billion.[4] It's worth noting that this investment in AI has also been massively concentrated in the United States. In 2022, the $47.4 billion invested in the US was three and a half times higher than its closest rival China, and ten times higher than its next closest rival, the UK.[5]

Generative AI has been heralded as a once-in-a-lifetime technological breakthrough that could affect everything from logistics to law, medicine and finance. Will generative AI live up to the hype? For investors, its long-term prospects as a real-world application are only a secondary consideration. So long as they pump up the stocks and valuations continue to soar, no questions are asked and the party continues. The buzz around AI provided jet fuel for the market's rise in 2023, with tech company financial reports buttressing the confidence of investors. Throughout 2023, Amazon, Microsoft and Alphabet's share prices were exceeding forecasts with strong profits and promises of future growth on the horizon. After getting pummelled in 2022 and suffering a roughly 60 per cent stock drop, even Meta was reaping the rewards from its cost-cutting and investments in AI, with a year to date gain for its stock of 55 per cent in late 2023.[6] Overall, the IT and communications sectors were up more than 40 per cent in 2023, with plenty of optimism that there was more AI-fuelled growth ahead.

Much to the annoyance of legacy tech companies that had been making cutting-edge advancements in AI for well over a decade, such as Google, Sam Altman, then CEO of OpenAI, emerged as the hero of the day and the man who would bring AI to the masses. By early 2023, this international ambassador of AI was on a global tour, meeting world leaders, venture capitalists and technologists, calling for 'global cooperation' to make safe and responsible AI, and helping to ensure that any future regulation did not impede his company too much. The entrepreneur was then just a man with a simple dream. As Altman himself jokingly remarked, 'AI will probably most likely lead to the end of the world, but in the meantime, there'll be great companies.'

Following the launch of ChatGPT, OpenAI was catapulted into the lead of a new generation of AI companies. In April 2023, it was reportedly valued at $29 billion, then in October it was in talks with investors for a valuation at $86 billion, roughly three times the price of only six months earlier.[7] While starting out as a non-profit, OpenAI transitioned to an unusual governance structure of a 'capped-profit' company in 2019 (investors, including VC funds, were limited to a 100x return), which was ultimately controlled by a non-profit entity that had a majority stake in the business. The idea behind this structure was to maintain the mission of developing AI 'for the good of humanity', with the pursuit of profit taking a back seat. Whatever good intentions may have led to this strange setup, such pretences are long gone, with the AI race to commercialisation in full swing. The scope of OpenAI's stated goal of striving to 'advance digital intelligence in the way that is most likely to benefit humanity as a whole' was quietly reduced to just benefitting Microsoft's shareholders. Pretty soon, the company had dropped its ban on the military use of its AI tools and was working with the Pentagon on cybersecurity software.[8]

OpenAI was joined by a collection of other even younger AI startups, which were all pulling in hundreds of millions of dollars in new capital. These companies include Anthropic, founded in 2021 by former employees of OpenAI, which received $124 million in Series A funding in its first year to help the company develop computationally intensive research on large-scale AI models. The company's chatbot Claude is a competitor with ChatGPT. Investors include Jaan Tallinn, co-founder of Skype, Eric Schmidt and others – $500 million of its funding had come from Alameda Research, the cryptocurrency trading firm set up by convicted fraudster Sam Bankman-Fried. As the funding suggests, Anthropic is close to the 'effective altruism' movement, a community that promotes doing the greatest good with your money, but which more recently has moved to a focus on the extinction risks posed by AI.[9] The *New York Times* reported that Anthropic was the centre of 'A.I. Doomerism', with employees comparing themselves to atomic bomb inventor Robert Oppenheimer.[10] In late 2022, Google invested $300 million in Anthropic for a 10 per cent share of the company, which also resulted in Google Cloud becoming Anthropic's preferred cloud provider. But then in September 2023, Amazon reported it was investing $4 billion in the startup, which would lead it to take a minority stake in the company and snatch up the position of its primary cloud computing service.[11] In addition, Anthropic agreed to use Amazon's specialised computer chips rather than Nvidia's for their foundation models.

Also in the mix was Cohere, founded in 2019 by two researchers who had worked on AI at Google, Aidan Gomez and Ivan Zhang, in addition to Canadian entrepreneur Nick Frosst. Aidan Gomez had been a co-author of the groundbreaking 2017 paper 'Attention Is All You Need', which paved the way for the latest generation of LLMs. Two years after its founding, Cohere raised $40 million in Series A funding, led by Index Ventures, with a promise to make natural language processing safe and responsible for business. In 2021, the former Google employees partnered with Google Cloud to provide the infrastructure for Cohere's platform. Cohere focused on generative AI for enterprises and sought to build tools for searching, summarising and writing that could be developed for businesses. For example, in 2023 it announced a collaboration with McKinsey to integrate AI products into the organisation's operations. These are just three of the most financially backed AI startups – by the end of 2023, there were over 200 AI companies valued at over $1 billion.[12] If there was a broader downturn in the markets, somebody forgot to tell generative AI.

Funding structures for new tech startups have been altered by changes in the global macroeconomic environment. Higher interest rates mean more expensive debt and less speculative cash floating about, which has produced a shift in Big Tech's developmental dynamics. Now, the high-valuation winners of the startup world are increasingly sheltering behind legacy tech companies, which provide them with substantial funding and support. This has also been influenced by the capital-intensive nature of AI development. As we saw in Chapter 3, the costs of computing power and using the latest infrastructure to train AI models is hugely expensive, requiring a level of financial backing that gives larger tech companies a degree of control over the direction of the startups. Sam Altman remarked that OpenAI might be 'the most capital-intensive startup in Silicon Valley history'.[13] In previous generations of Silicon Valley companies, angel investors could have been friends, family and other entrepreneurs who had made it big during the dot-com boom. In 1994, Jeff Bezos received $300,000 in startup capital from his parents and then convinced twenty-two generous friends to invest $50,000 each into his idea for an online bookshop.[14] Mark Zuckerberg received a $500,000 angel investment from venture capitalist Peter Thiel (more on him later) in exchange for a 10.2 per cent stake in the company. Meanwhile, even Jeff Bezos got in early on Google by making a $250,000 investment into the fledgling search engine in 1998 – an investment which if he still held would now be worth $1.5 billion. By 2020, if you were a young AI startup and wanted to join the club, you had to partner up with one of the older kids in order to play the game. Legacy tech companies were among the most liquid on the market, with enormous cash reserves and a desire to invest in startups with good people and ideas. Amazon, Microsoft and Google have spent billions on startup AI companies, but they are also charging them an undisclosed amount to use their cloud platforms.

There are signs that many AI business models may actually be able to generate significant revenue, or even profit. For the Silicon Valley VCs, this really is just an added bonus. As mentioned above, they predominantly make their money when startups are acquired by bigger firms or make it to Initial Public Offerings (IPOs). They care about telling a story of growth and disruption that justifies hopes of future profitability more than the long-term viability of the business. But even if only a fraction of the startups work out, the investors will have long since made off with their bags of loot. The high valuations of AI firms promise a lot – and, ultimately, it's the promise that matters to early stage investors. In terms of precisely how these AI companies will bring in revenue, it's still very early days for generative AI and there are likely a number of different business models that are yet to be developed. Early models have focused on subscriptions, pay per usage and integrating AI products into an existing business. At the same time, current valuations of many of these startups are extreme, even by Silicon Valley standards. To understand how the decisions of people like Tyler shape the development of technology and the lives of people implicated in the movements of the extraction machine, we need to understand the longer history of capital. The capital that Tyler invests has its own family lineage: every dollar contains within it cents that come from previous investments, and if we follow the cycle back over and over we can trace how the extraction machine first emerged.

The Making of Californian Venture Capital

The name 'venture capital' is relatively new, emerging in California in the early 1960s, but the original concept has been around for millennia. Ever since merchants were filling long-distance trading ships with spices, sugar, ivory and slaves, the rich have ventured their money to try and make more money. Every society with functioning trade networks and the exchange of commodities for money has had its own class of rich people who buy low to sell high. But capitalism didn't start in the docks and wharves of trading cities – it started in the fields.

We have to go back to a very specific time and place: capital emerged specifically in the agriculture of medieval England. Feudal lords used their political power to take economic surpluses from peasants. But they didn't have total, dictatorial control of everything. They had to abide by a system of law called manorial custom. These customs regulated lots of things, including land use: they embodied, according to historian George Comninel, 'historically developed expressions of communities working out effective means of living and producing together'.[15] Custom gave peasants common rights to land they didn't own. It also forced the lords to submit to collective decisions on questions like what crops should be planted and when fields should lie fallow. These rules meant that the local economy was always subject to a web of collective rights that gave ordinary people a voice. Enclosure was the process that undid this system of regulation. In short, it meant the lords taking exclusive control of their land. They applied new agricultural techniques and made the land more productive. The word 'enclosure' sounds very literal, like the main element of the process was fencing off land from the peasants for the lord's exclusive use. But enclosure was primarily about eliminating a system of economic regulation that gave ordinary people a say and replacing it with a form of law based on private property. The fences were just a side effect. The people who owned the land got to make all the decisions, and the people who worked on it had to do what they were told. Private ownership of the means of production creates a situation where non-owners (be they peasants or workers) have no democratic right over how the economy operates. Capitalism has been based on this principle since its very beginnings.

Enclosure was the start of a transition to agricultural capitalism in England. The elimination of the common rights to land created a class of people with nothing to sell but their labour, who would later be pulled (sometimes literally) into the first extraction machines of the modern era in the forms of the spinning jenny and the steam engine.[16] The regime of singular control of production, first created during enclosure, would come to be the norm across the whole of society: in the factory, there was never any system of collective regulation. The owner was always, exclusively, in charge.

Capitalism's unquenchable thirst for new territory to enclose meant that it could not be confined to England's hills. It spread. It took hundreds of years to arrive in California, but when it did, it did so with ferocity. Here, in the warm sun of the Pacific coast, enclosure arrived accompanied by settlement and genocide. When European settlers first arrived, the population of the Indigenous peoples of California has been estimated at close to 300,000.[17] By the time the US Pacific Squadron dropped anchor at Monterey Bay and declared California US territory in 1846, that number had fallen to about 150,000, a steep decline of 50 per cent over the course of just eighty years.[18] What would happen next, however, would be even worse, with the violence involved shaping the specific form of Californian capital. In the twenty-four years of the 'population cataclysm' that followed, California's Indigenous population would be subject to what historian Benjamin Madley describes as an 'American genocide'. He focused on how settlers' racism and fears of the Indigenous population combined with their desire for land and resources. This resulted in a settler-led process of killings with extermination as the goal.[19] While the settler population of California rose from under 10,000 in 1840 to a total population of 864,694 in the 1880 census, the Indigenous population fell by 80 per cent to around 30,000 in 1869, before declining further to 16,277 by 1880. Population declines this profound are hard to find comparisons for, even in the absolute worst moments of the twentieth century. Death came for the Indigenous Californians in many ways: the more indirect forms of disease, dislocation and starvation combining with the more obvious causes of slavery, imprisonment, murder and massacre. The reservations that many Indigenous Californians were forced onto were, in the words of Mahmood Mamdani: 'a milder version of the Nazi's industrial concentration camp – a place of internment and slow death', run by military administrators whose stated intent was extermination.[20]

One of the fundamental issues at stake was exclusive ownership and control of land. The Indigenous population of California had their own systems of customary, collective land regulation, like the peasants of England. As complex hunter-gatherers, their lifestyle sat somewhere between mobile hunter-gatherers and more settled agrarian societies. They built sedentary villages, reached high population densities and created complex social organisations, but never built their economies around growing corn, beans and squash. Instead, they developed an 'economy of diversification' predicated on forms of landscape management that enhanced resource diversity and availability.[21] Such an approach was fundamentally incompatible with the settler logic that claimed exclusive possession and control of natural resources. The two pillars of capitalist development in California after the gold strike in 1849 were resource extraction and agriculture, both of which implied claims on land and water that needed to go uncontested if capital was to grow to its gargantuan potential. But whereas enclosure in England had upended peasant control of the land and created a massive impoverished workforce that could be integrated into the dark satanic mills of early industrial England, enclosure in

California aimed at genocide. The Indigenous peoples of California were dispossessed without being proletarianised, and eliminated rather than exploited.[22]

The enclosed land left behind after the genocide became a prime resource for the development of a specifically Californian capitalism. In the wake of the gold rush came an agricultural boom. By 1939, lawyer Carey McWilliams was writing of a 'a new type of agriculture . . . large-scale, intensive, diversified, mechanized'.[23] This agricultural sector exported around a billion dollars of fruit and vegetables a year, and was capable of providing for the needs of the whole nation. Roughly 250,000 migrants worked the fields, and whenever they organised they were faced with a system of repression so extreme that McWilliams called it 'farm fascism'. This workforce was produced by a system of state and federal immigration controls which at various times excluded Chinese migrants (1882), restricted the immigration of Japanese labourers (1907), banned any migrants from a 'barred Asiatic zone' that spanned from Afghanistan to the Pacific (1917), created explicit racial quotas for migration (1924), and launched a violent 'repatriation movement' based around mass deportations (1931).[24] All of these attempts at control struggled with a fundamental contradiction. The state was both reliant on and intolerant of migrant agricultural workers. As Malcolm Harris puts it in his brilliant history of Silicon Valley, Palo Alto: 'California's agricultural capitalists pedalled the state's nonwhite labour like a bicycle: When they pushed one group down, another rose to replace it, and the whole contraption moved a little further down the road.'[25]

During this development, the foundations were laid for what would become the cornerstone institutions of Californian capital. It was the wealth created by an agricultural boom and a gold rush that went on to become the deposits in San Francisco's own Bank of Italy, founded in 1904 and eventually renamed as the Bank of America in 1930. Meanwhile, in Palo Alto, a horse ranch set up by a railway magnate keen to escape the angry workers of urban San Francisco metastasised into Stanford University. The university became an early base for radio technology developments (alongside its other preeminent field, eugenic race science).[26] When the US entered the Second World War in 1941, the flood of federal military funding played a vital role in turning the university into the centre of technological development it is today. What Senator J. William Fulbright termed the 'military-industrial-academic complex' had one of its preeminent bases at Stanford.[27] Research in electronic engineering, radio and radar was the bread and butter of the university's work during and after the war.

The Soviet launch of the Sputnik satellite in 1957 accelerated the federal government's investment in research, as universities like Stanford began to be seen as a key force in the Cold War. Over the decade between 1955 and 1965, federal research funding climbed from $286 million to $1.6 billion through mechanisms like the National Defense Education Act, until the Department of Defense was the biggest research funder in the country.[28] This investment was particularly concentrated in California, where a golden triangle of academic, military and industrial science started to churn out innovations at a world-leading rate, from the silicon semiconductors that gave the Valley its name to the other technologies of modern computing. The American government was, as historian Margret O'Mara put it, 'the Valley's first, and perhaps greatest, venture capitalist'.[29]

But the venture capital industry as we know it today wasn't far behind. Big wealth funds had always made some risky investments in small companies, but VC increasingly emerged as a specific form of early stage financing to buttress the development of the electronics industry.

The first VC fund in Palo Alto was founded in 1959 with $6 million in capital, and before long it was joined by others.[30] Again, the federal government played a vital role in enabling this growth. The Small Business Investment Act gave VCs big tax breaks and access to federal loans that supersized their capital funds by 300 per cent.[31] By 1961, there were over 500 Small Business Investment Companies (SBICs) around the country.

The particular innovation of VC investors in California, however, was that they adopted a specific approach to investment. Unlike East Coast VCs, who tended to take minority stakes and have a relatively hands-off relationship with the firms they invested in, Palo Alto VCs went all in. They adopted a limited partnership model, which meant they secured a large percentage of the firms they invested in and developed a close relationship with the founders, often guiding them through key decisions and processes. They inserted themselves as decision-makers into the process of technological development at an early stage, thereby aggregating huge private decision-making power to direct the outcomes of publicly supported research.[32]

The model worked, and business boomed – but only temporarily. VC went into a kind of hiatus while the economic headwinds of the 1970s blew, as the risks involved in small electronics investments increased beyond the appetite of institutional investors. This meant that VC firms struggled to find anyone to sell their stake to, and valuations were lower. The hiatus would gradually end, however, and then be fully reversed by the burst of innovation that surrounded the development of the personal computer and the Internet. New capital commitments to VC firms rose by 2,000 per cent between 1991 and 2000, with most of the cash coming from pension funds.[33] The average rate of return for funds rose from the prolonged lows of the 1980s to over 100 per cent by 1999. VC's success list included Apple, Microsoft, Cisco Systems, and Sun Microsystems.[34] The dot-com crash would take the shine off this incredible burst of value creation, but the VC industry would recover and come back stronger than ever before, as if the inflation of a colossal asset bubble had been a freak accident rather than a symptom of underlying systemic instability.

A huge proportion of the research that made this accelerated development possible was funded by public money. Silicon Valley was therefore always a product of both Californian capitalism and the American state apparatus in a broader sense. From the National Defense Education Act to the Small Business Investment Act, the institutions of Californian capital needed the support of a particular legislative agenda in order to become what they are today.[35] Silicon Valley was founded on government research dollars, and supplemented by tax breaks and federal loans. Capital's putative independence was only skin deep. Despite this public contribution to the foundational research that created the conditions for the tech industry, VCs were never under any democratic obligations to direct their investment for the public good. They even fought battles to defend the enormous profits they made against capital gains tax, meaning that a greater share of the commercial benefits of public investment became accumulated by private actors.

Even when funded by public investment, the decisions that have shaped the course of technological development for the past century have been taken mainly behind closed doors. Ordinary people have been the subjects of technological change, rather than the agents shaping it. Today we live in an economy that is characterised by private profits and private decision-making: the corporations shaping our future are, in a phrase attributed to the famous linguist Noam Chomsky, 'islands of tyranny', and their foundations were laid on the basis of particular histories of genocide, racism and exploitation.

Technology Without Democracy

Technological development can sometimes seem like a natural process. You might be familiar with Moore's law: the idea that the power of computers doubles roughly every two years. Its description as a law makes it sound fundamental, like Newton's laws of motion. It's not much of a stretch to imagine that all of human history has been subject to a version of that law. Each eon sees the next phase in a long, linear process of development from the wheel to the large language model, via the printing press and the mobile phone. Only that isn't how technology works at all. The rules of its development are, in fact, entirely unlike the laws that governed the apple that fell onto Isaac Newton's head. Technology is a social force, and its rules are the same ones that shape our everyday lives: rules of profit and growth, expansion and domination. To better understand technology, we should examine the systems and conflicts that shape it. In the preceding two sections, we've seen the contemporary business context that AI companies and their investors are operating within: the new era of AI. We've also seen how the history of capital generally, and Californian capital specifically, has created both the exclusive right of owners and managers to dictate how technology develops, and the pool of concentrated capital that has become Silicon Valley. Now we will zoom out to address our final two questions: which structural pressures are shaping the technological development of AI; and why the worldview of tech executives concerns us.

The primary drivers of the development of AI are large tech firms operating in a competitive and globalised capitalist economy. Competition between these firms means they are always under pressure to lower the costs of production. In practice, this means they either need to squeeze more output from workers (by intensifying production or reducing the wage bill) or invest in new technology that increases productivity or eliminates waste. The development of new technology in this context gives businesses an advantage over their rivals and encourages them to be constantly in search of new innovations. Competitive market pressure between firms provides the structural constraints within which individual companies must operate.

Deciding how to navigate these pressures is the job of the people who own and/or act on behalf of capital: a category that includes a whole range of managers, bosses, shareholders and investors. In making these decisions, they are rarely under any kind of obligation to share agency with workers. In the vast majority of companies in the world, most members of the organisation have little to no ability to shape its operation. As a result, the needs and desires of an elite clique of managers and shareholders are prioritised above the needs of the workers. Few actors within the global production networks of AI have access to the power to fundamentally reorganise the relationships they find themselves trapped in. Most of the stories we have examined so far have seen their protagonists struggling to exercise agency in their daily lives. However, some actors within these networks have privileged access to power, control and knowledge. It is at these points, in the boardrooms where investors and owners gather, that the decisions which shape the network are made.

Which worldviews inform the perspectives of decision-makers in these spaces? Silicon Valley is well-known for its long history of a distinctive mixture of libertarian politics and neoliberal economics that was succinctly captured almost thirty years ago by Richard Barbrook and Andy Cameron's idea of a 'Californian Ideology'.[36] This ideology was a blend of ideas from the New Left and New Right, combining a romantic individualism from the counter-culture Left with the anti-statism and free-market economics of conservatives. Added to the mix was a technological determinism and optimistic belief in the liberating power of networked personal computers. While elements of this still ring true today, certain things have changed over the past decades.

Since this time, many tech workers in Silicon Valley have become more socially progressive and critical of the hyper-individualistic and techno-utopian aspects of tech culture, with growing concerns for racialised and gendered forms of oppression and the impacts of the climate crisis. The Californian Ideology predominated at a time when it could be interpreted as a vague belief in how the spread of technology and information would drive wealth creation and human freedom. Yet as the 2010s unfolded, the problems of new forms of surveillance, tech monopolies and algorithmic discrimination became more apparent. The polarisation of American politics that accompanied Donald Trump's electoral rise also forced the tech world to make more explicit partisan commitments. Some have become mega-donors of major parties, and supported prominent social movements such as Black Lives Matter and Me Too.[37] It is undoubted that most tech executives lean progressive and support Democratic candidates and causes, but it would be misleading to suggest they all adhere to a single political perspective. The tech world has both progressive and conservative sides to it and a wide plurality of differences in between.

There is, however, an underlying agreement about who should be making fundamental decisions about how technology is controlled and developed. We call this *the founder's mentality* – the idea that startup founders and tech CEOs are best placed to make the most important decisions about which technologies receive funding and how they should be built. It's a belief in markets over democracy, in business and philanthropy over government spending, and in industry self-regulation over democratic laws.

Founders see themselves as the most important drivers of change, and democratic processes as a necessary yet secondary element that creates legal frameworks and tax incentives, but which should not overstretch with restrictive regulations on business. Even those CEOs who speak publicly in favour of government regulation often bristle at instances when it negatively impacts their bottom line.[38] This founder's mentality could be considered what political scientists call a 'thin ideology' – a worldview that only addresses part of a political agenda and is usually attached to a more comprehensive host ideology such as liberalism or conservatism.[39] It therefore comes in different shades, from the socially progressive to the libertarian conservative. But regardless of their political persuasion, most tech executives believe that at the end of the day they are the ones who should be calling the shots.

At the progressive end of the spectrum, you have tech CEOs such as Mark Zuckerberg, who supports a host of progressive causes and has established a major philanthropic organisation. Yet despite these progressive tendencies, Zuckerberg runs Meta as his personal fiefdom, commanding 61.1 per cent of the voting power due to the unique voting structure he established at the company.[40] This gives him the power to configure what Facebook and Instagram's algorithms allow people to see, what posts get deleted from the platforms and which competitors to buy up or copy. A single human being has the power to manipulate the digital infrastructure used by billions of people across the world.

When asked about the power billionaires exercise in American society at an internal Facebook Q&A, Zuckerberg responded as follows:

> There shouldn't be an accumulation of private wealth that allows people to . . . we're funding science for example . . . some people would say is it fair that a group of wealthy people get to, to some degree, choose which science projects get worked on . . . I don't know how to answer that exactly. At some level it's not fair, but it may be optimal, or better than the alternative. The alternative would be the government chooses all of the funding for all of the stuff.[41]

Tech executives like Zuckerberg want us to trust them. Despite the blatant unfairness in this huge disparity of power, Zuckerberg falls back upon the implicit argument that it's all for the greater good. For him, ultimately, savvy entrepreneurs and private actors are better placed to make important decisions than democratically elected governments. The tech industry operates as a kind of meritocratic filter, and the people who rise to the top will tend to make the right call more often than not.[42] Tech executives like Zuckerberg assure us that they have our best interests at heart. Meta's mission is 'to give people the power to build community and bring the world closer together'. What's not to like? He sees himself as one of the good guys: he wants Meta to be more inclusive, have a higher percentage of racially diverse staff, for LGBTQIA+ people to feel empowered and for technology to benefit humanity as a whole. Yet all of this sits comfortably beside his belief in the unlimited private accumulation of wealth and the exercise of private power over tech development. Ultimately, it's the entrepreneurs and builders who should be in charge of shaping the infrastructure and tools that power modern life.

The problem with this founder's mentality is that it does not sensitise CEOs like Zuckerberg and others to how dangerous their own unchecked power can be. Founders do not see the world as one of competing interests and potentially antagonistic social groups. Instead, the world is full of technical problems to be solved and smart people that just need the right funding and support. Zuckerberg and his ilk believe the world is generally getting better and they are playing a key role in making this happen. They are long-term optimists and still believe in the emancipatory power of technology. They typically invest a lot of their firm's time and money in 'tech for good' initiatives without asking the messy political questions of whether 'good' for one group is a harm for another. Because they don't see themselves as a threat, they have no problems exercising near unparalleled power over our lives.

This is a view that Zuckerberg shares with another Silicon Valley founder, multi-billionaire venture capitalist and Godfather of the PayPal mafia,[43] Peter Thiel. As a self-styled conservative libertarian, he sits on the more right-wing end of the Silicon Valley political spectrum, with more extreme views of how far the founder's mentality should be taken. Thiel believes in the total elimination of government regulation in order to unleash a very particular vision of human freedom based around private property and the absolute rule of the market.[44] It's a vision that builds on the most aggressive elements of the neoliberal doctrine advanced by twentieth-century thinkers such as Ludwig von Mises, Wilhelm Röpke and Friedrich Hayek.[45] And just like the neoliberals who came before him, Thiel's vision of human freedom is based on an unlimited freedom to profit (and by implication, to exploit), rather than a freedom to participate, decide or enjoy. Although Thiel's substantive political philosophy differs significantly from many other tech executives, there is a shared belief in the importance of their own position in exercising a controlling influence over the development of new technology.

Thiel's pursuit of this vision is already shaping the use of AI on a huge scale. Thiel is a founder and current chairperson of the board at Palantir, a global surveillance company with a market capitalisation of $38 billion. Between 2007 and 2021, the company held 940 contracts with US federal intelligence and security agencies worth over $1.5 billion. The company offers what critics have described as 'surveillance as a service' to the US police and military. Media studies scholars Andrew Iliadis and Amelia Acker state this includes 'technologies for data gathering, labelling and modelling in contexts that include facial recognition, predictive policing, workplace surveillance, and social media mining' with the goal of 'assist[ing] clients in amassing a shadow data infrastructure for purposes of governing and control'.[46]

Thiel went on to fund Clearview AI, a facial recognition tool that uses three billion images scraped from the internet to help immigration enforcement identify and detain people for deportation and which was subsequently rolled out to over 600 police departments. Alongside Safe Graph (a real-time database of human location data) and Anduril (building smart border infrastructure), these projects make up the Thielverse: a set of AI-enabled data and infrastructure companies which profit hugely from facilitating the expansion of the American surveillance state. The commanding influence of these overlapping institutions is explicitly guided by a philosophy based on anti-democratic elite control.

The founder's mentality can take a variety of forms – from the family friendly global community builder who just wants you to connect more with your friends to the anti-democratic shadowy architect of a mass surveillance system. Yet despite these differences in political perspective, there is something that unites many of the tech executives of Silicon Valley. They share a suspicion towards the role of government and an innate belief in their own good intentions. Tech executives position themselves as self-designated thought leaders of the global community. It is their trillion-dollar companies and elite philanthropic organisations that are changing the world, and cumbersome processes like democratic voting and public policy tend to get in the way.

Under the watchful eyes of these founders, artificial intelligence has taken on a particular guise. It could have been – and could still be – many things: a work-eliminator, a resource-planner, a human-emancipator. But instead, it is becoming a work-accelerator and a surveillance-intensifier. This path has been shaped both by the structural pressures created by the dominant global economic system, and by the political visions of people with privileged access to power, control and knowledge. Technological development is under the control of the most powerful tech companies with strong imperatives to continue expanding their operations and increasing their profits. When these actors are the ones in a position to control how AI is developed, it should be no surprise that the products they consider meaningful and important are those most profitable to other companies. Whereas most people entangled in the extraction machine have little say in its overall operation, others have the almost unique ability to make profound, world-changing decisions about its future path.

16th Street and Mission

Tyler emerges from an Uber in the Mission district of San Francisco. He's meeting four of his old NGO friends for dinner. He doesn't come down here much anymore: the city's intensifying homelessness crisis is more evident in some places than others. Here, by the BART metro station, you can see it. But more than seeing things – tents and wheelchairs and empty bottles – you can smell them. It is a warm evening. A man's naked foot is poking out from under a dirty blanket, twitching irregularly. The five friends walk to a small Taqueria, drink some Modelos and tell each other stories about their jobs and the assholes they work with. The nostalgia kicks in, and they start trading jokes about when they had just left college and were first working out how they could make a difference. They walk up to the benches at the top of Dolores Park on 20th Street. In the night sky, they can see the irregular teeth of the city's skyscrapers, their lights acting as a particular point of emphasis with the more general backdrop of light pollution. Above them are a few stars and a few planes. It's a full moon.

Earlier today, Tyler decided to strongly suggest that the health and safety startup's founders relocate their data annotation from Alabama to East Africa – and when he strongly suggested things to founders, they usually happened. This was the job of a venture capitalist, after all – to draw on their experience and knowledge to make the key decisions that shaped the trajectory of the company and its product. The potential savings are huge, over 70 per cent. If they are realised, it will massively improve gross margins and send the numbers overall much further into the green. The founders are having a meeting with a few potential annotation firms next week to discuss what the new service would look like. They're anxious to make sure the annotation is of the same quality, but Tyler feels confident it will be. He earlier felt a persistent twinge of concern about the impacts of the relocation on the annotators based in Birmingham, Alabama, but now kept returning to one central question: wouldn't it be even more unfair to deny people in the Global South jobs just because wages are lower in their economies? Someone was always going to lose in a global economy; this time it was the people of Birmingham. Rather than taking work away from Alabama, he thought of himself as giving it to Kenyans and Ugandans. He had ended up scrolling through the social-impact sections of potential providers earlier, looking at the smiling faces of data annotators who were making better lives for themselves and their families through access to work.

Those annotators might have nothing in common with him, but

they would be working on the same project. That was the wonder of the global division of labour. He thought about the cloud services they relied on, the server racks humming quietly in whatever convenient location their provider had thought to stash them. About the designers putting the finishing touches on the new user interface. About the workers pushing trolleys around a warehouse under the benevolent gaze of the system. About the engineers, based somewhere in the city below him, working out how to bring all those inputs together into the finished thing. His job changed things. It might not be simple, but it changed the world. And that's what he always wanted to do.

7

The Organiser

The spark is lit during a routine meeting between staff and management in a dull office block on the outskirts of Nairobi. Paul is frustrated and upset by the lack of support given to him and his colleagues. 'We're being asked to do too much! Please give us more time for each decision. We are being asked to look at very traumatic videos.' Paul is part of a team of content moderators in the same centre as Mercy, who we met in the Introduction. They work under appalling conditions involving short-term contracts, long shifts, relentless exposure to shocking content and mind-numbing repetitive work. By this point, Paul is seething with frustration. He has had enough of management's excuses and knows the company could do better. Despite the seeming futility, he firmly tells the manager once more just how much he and his colleagues are suffering. He is at breaking point. This time it has to be different.

The manager chairing the meeting stares back at him blankly. Rather than engage with the substance of what Paul is saying, she launches into a long monologue about key performance indicators and the company's precarious contracts with its foreign clients. She sometimes drones on like this for ten minutes without stopping, talking to them like they were naughty schoolchildren who needed to be taught the sad realities of the big wide world. The message is clear: 'We can't afford for you to slow down – these big tech companies will move your jobs to India if we don't deliver.' This is how management typically deals with such concerns: some corporate drivel that results in inaction.

But, this time, something unexpected happens. Other workers respond. Voices are raised. The tension in the room is palpable. It's not just Paul who's had enough – his colleagues are furious. Others start to detail the litany of issues they face at the firm and make demands for how things should change. Many of those speaking up know the risk they are taking. Being labelled as a troublemaker means your short-term contract might simply not be renewed after it is finished. Some of the workers suspect that management has 'informants' among the employee body. Whether or not this was in fact true, management had a very low threshold for any kind of resistance.

The manager looks wide-eyed around the room. She shuffles the paper on the desk in front of her and pretends to make some notes while the team complains. She is doing her best to disguise the fact that she is rattled. This has never happened before. As soon as she can call the meeting to a close, she bolts for the door. The response is swift. Management decides to stop having physical meetings and instead hold virtual ones where staff are no longer given the opportunity to ask difficult questions. None of the concerns raised by workers are addressed. The unrest at the previous meeting is attributed to a few key troublemakers who are kept under close watch.

But instead of silencing the workers, this decision has the opposite effect. No longer able to speak up, they start to organise. It begins in a WhatsApp group. Paul invites a select group of colleagues and together they start to discuss all the problems they face. Grievances are aired; jokes are told; memes are shared. The group quickly develops a shared sense of how badly they have been treated and that this is a structural feature of their workplace. Over time, Paul and the group start to organise in-person meetings to discuss and strategise. The meetings are small to begin with. Some colleagues feel relieved and empowered that there is now, finally, open discussion about the mistreatment they endure. Others are scared. They have internalised the message from management about how precarious the company's own contracts are with its clients and are worried about the company losing vital work.

Every worker who attends these meetings knows they are taking a risk and that collective action is a perilous path for them and their colleagues. After weeks of back and forth, the group decides to create a petition with a list of demands:

> We demand equal pay with the rest of the world in regards to content moderation. We have to demand for more to be done in mental health, better pay and better working conditions. Content moderators are vital contributors to the digital landscape, and their well-being should be a priority. By addressing these issues, we can create a healthier and more productive workplace environment.

Almost 200 workers sign and put management on notice that if their demands are not met within a week they will resign from the company.

Rather than accede to the demands, the company fires Paul – accusing him of bullying his colleagues. They then send in senior managers to talk with the rest of the group. Some are reminded just how many people are willing to take their jobs; others are enticed away with promotions if they help management persuade their colleagues to get back to work. There is still unrest, but the company tries to dismantle the core group of organisers.

Paul contacts a few journalists from the international press, inviting them to report on the conditions at the firm. Following an initial exposé, multiple pieces are published, all revealing the shocking details about the conditions experienced by data workers in Nairobi. The organisers hope that this pressure will force the company to address their concerns.

Senior management has other ideas. The content moderation contract is becoming too much trouble, and with the media attention, it has started to become a serious reputational liability for the firm. The company pulls out of the contract with the social media company and decides to focus on data annotation. Hundreds of content moderators are laid off. There are no winners: the client offers the contract to another data annotation firm with a large presence in Nairobi known for even worse pay and conditions.

A protest is organised outside the office and an unfair dismissal lawsuit is brought against the social media firm and the two Nairobi-based outsourcing firms. Posting on the WhatsApp group, one of the organisers writes: 'One of the reasons we would like to protest is because our employer doesn't like their reputation being tampered with. They want the world to know that they are ethical. But we would like the world to see the situation we are in.'

On the steps outside of the building, the crowd are holding print-outs of the petition – waving them in the air. 'We are going nowhere!' shouts one of the ringleaders in the crowd. 'Yes!' comes the response. 'We are going nowhere!' another protester shouts. In union, the crowd responds: 'Yes!' After the day-long sit-in, the group is eventually dispersed, but this is not the end of the story.

The organisers realise that without a formal trade union, they have no legal mandate or right to strike. Few of the workers have ever been a member of a trade union and are not sure where to begin. Paul had initially looked into the documentation for union incorporation before he was fired and passed this paperwork to his friends. Then, on International Workers' Day 2023 (1 May), three days after the protest, the African Content Moderators Union (ACMU) is voted into existence by data workers at a packed room at the Mövenpick Hotel in Nairobi. The 150 workers who had gathered to form the union are jubilant. Confetti falls onto the stage and music plays as the crowd claps and cheers in celebration. As a legal entity under Kenyan law, the union will help future generations of data workers in their struggles for decent and dignified work. Paul would never get his job back, but his legacy as an organiser is now permanently enshrined in Africa's first union for content moderators. It will be a crucial institution to ensure workers have a collective voice and can take action to struggle for a better future.

Can Data Workers Organise?

What are the prospects for the African Content Moderators Union and other similar collective organisations of data workers across the world? How likely are they to achieve their goals of improving pay and conditions in their sector and giving data workers the respect and dignity they deserve? In late 2023, we spoke to three of the ACMU organisers – Rose, Mohammed and Brian – about what had been achieved in the six months since the formal incorporation of the union. The ACMU had been established by current and former content moderators who all worked at outsourcing companies on services for Meta, ByteDance (TikTok) and OpenAI. We were hopeful, although a little tentative given the enormous forces amassed against them. It's worth recalling, after all, that the ACMU is attempting to negotiate with some of the largest tech companies in the world and their outsourcing partners.

Rose told us that the vision of the ACMU was for content moderation 'to be recognised as a skilled job in which you have to get certified'. She went on to describe the need for more government regulation to ensure minimum working conditions and decent wages. But despite the enthusiasm felt by all on the day of the union's founding, some things had not progressed well in the following six months. The actual conditions and pay of content moderators in Nairobi had not improved. If anything, they had gotten worse as the content moderation contract had initially passed to another company that was even less concerned with ensuring decent work for its employees. Things went from bad to worse when this second company then lost its contract with Meta, leaving hundreds of Nairobi-based workers facing the threat of being laid off. Many of these workers support their families with their income and have found it difficult to transfer their content moderation experience into other sectors of the local labour market.

This is not surprising, since a lot of the tasks assigned to workers require very little training. That is not to say that jobs in data work are simple, or that a significant amount of learned proficiency does not go into doing the work with speed and precision. Rather, these are jobs that are designed in such a way that human creativity and individual skill is largely bled out of the system. Workers are integrated into a system of production in a way that means their outputs have to be standardised, which leads to both routinisation and deskilling. The human data workers are effectively part of the company's algorithm.

Workers who leave their roles do not take with them large investments in human capital.

The ACMU organisers were also concerned about low wages. One of the founding principles of the union was the demand for equal pay with the rest of the world. Kenyan data workers produce an immense amount of value for firms based in the Global North, and yet they see only a sliver of that value returned to them. However, this demand is a particularly difficult one to reconcile with the structure of the planetary market of data work. They must compete against workers from every corner of the globe, which depresses wages across the sector. While content moderators require a degree of cultural and linguistic familiarity with the places their data work will be used, East Africa is not the only region with moderators who have those skills. Other data work also tends to neither require those skills or any proximity to the client – as far as they are concerned, the data workers might as well be on the moon.

Another problem for the ACMU was that despite the low wages and high-stress nature of their jobs, high unemployment in their country meant that there was a large group of people ready to step into their roles, which reduced their bargaining position. Content moderation tended to be higher paid than working in the informal sector, so data workers were reluctant to risk antagonising their bosses. Both workers and bosses were intimately aware of these structural dynamics, which created a situation in which managers and workers openly discussed these issues: 'We couldn't possibly raise wages; we have to compete with India.' 'The client is demanding this work by Monday morning. Our contract depends on it. So, we'll all have to come in on the weekend.' 'If you can't do this job, just leave it and another person will come and do it.' Rose recalled managers frequently telling workers who complained all these things.

What exactly could the ACMU do? When asked about the union's power to achieve its demands, and how it might exert this in the context of the planetary labour market for data work, the conversation went from hope to despair. Brian went first:

'The jobs come to us because we are cheap labour.'

'They can decide to find a cheaper option.'

It seemed there were few practical options for what could be done and the workers seemed despondent: 'Seriously, at this point, I have no specific idea. There is a rumour of them hiring in Ghana. They are trying to start afresh.'

Mohammed agreed: 'There is little to nothing the workers can do when companies decide [to move]; they can decide to move anywhere . . . Policies are just there to protect industry. They put the workers at risk.'

Part of the problem is that the ACMU does not have a clear guide for how it should proceed. The organisers have no experience with any other union of content moderators, and none of the three organisers we spoke with had ever been in a trade union before. This lack of knowledge about historical precedents of successful campaigns puts organisers like Rose, Mohammed and Brian at a further disadvantage.[1] Unlike in other industries with long histories of unionisation, such as train drivers or dockworkers, data workers have few examples to follow, which can leave the entire sector normalising the status quo.

When the ACMU started out, they had to build workers' power from the ground up. Established Kenyan unions were not a lot of help: 'Most of them don't even know what we're doing. They mistake us for content creators,' said Rose. When they heard the word 'content', most trade unionists assumed that the ACMU was a union of people doing dances on TikTok. Building their own organisation was really their only option. The ACMU received limited support from some international allies, such as a global trade union confederation. However, with most of the organisers volunteering and working unpaid for the union, they must develop their strategy in what little time remains in their busy days.

The case of the ACMU illustrates the problems faced by data workers with little bargaining power. Jobs are deskilled and low-paid, work is traded in a planetary market, workers are mainly non-unionised, and positive examples of successful union organising are few and far between. All considered, the formation of an organisation like the ACMU is particularly remarkable. Individual workers need to unionise if they want to push for better conditions. But, as we have seen, while organising might be a necessary condition for workers to improve their jobs, it is by no means a sufficient one. How, then, are other groups of data workers likely to fare?

When companies perceive workers to have limited bargaining power in the labour market, trade unions are left with two primary strategies to enact change. We call these *blocking the flow* and *sounding the alarm*. According to the first strategy, workers can organise a collective withdrawal of labour through a strike or 'going slow', involving slowing down the pace of work. This draws on their power as an essential part of a larger network and is more effective when workers are either difficult to replace or can prevent the network from operating. The second strategy refers to workers' ability to create public pressure for change through a 'name and shame' campaign which attacks the reputation of the firm. This approach attempts to change the behaviour of a company through the threat of negative publicity and forcing it to live up to its own ideals.[2]

The first of these strategies – collectively withdrawing their labour power – is a tried and tested method for workers to improve their pay and conditions. It generally works best when workers occupy strategic bottlenecks in production networks or economic systems. There are two main ways workers can block the flow based on their position in the network. We could say that a *topological bottleneck* in a network is a position of power due to where it is situated in the overall shape of the network. For example, if all goods or services have to pass through a specific place or set of personnel. Dockworkers are an example of this. It is hard to get large amounts of goods in or out of most countries without going through a port. Air traffic controllers, customs officials and lorry drivers can exert similar power in their networks, with the threat of immediate economic chaos and empty shelves if they go on strike. But a bottleneck can also be *temporal* when all work has to be done, or all goods have to be delivered, within a certain timeframe. If doctors or nurses go on strike, people may die in the meantime. For data workers to be able to successfully collectively withdraw their labour, and extract meaningful concessions, we need to ask if they have the ability to command any *topological* or *temporal* bottlenecks in their networks.

With regards to the first question, the answer is yes. Their work is essential and indispensable in all the production networks that they are part of. Social media platforms would not function effectively without content moderators constantly reviewing all flagged content; large language models would not function without workers validating outputs; autonomous vehicles would not be able to drive without workers who train them by annotating and classifying objects in the streets that they drive on.

When it comes to temporal bottlenecks, the answer is also yes in certain circumstances: some data is live. In Vietnam, we interviewed one outsourcing company that is embedded into the workflow of a large European postal service. Because the client firm in Europe needs to handle approximately ten million letters a day, they use local automated systems to quickly sort letters into local, regional, national and international piles. An optical camera scans text on the front of each letter and accurately processes about 80 per cent automatically. The remaining 20 per cent are letters – many, perhaps, with messy handwriting; some containing multiple addresses – that need a human worker to decipher and make a judgement about which pile they belong in. In those cases, a scan is transmitted to one of 500 workers' screens in an outsourcing centre in Ho Chi Minh City. That worker has less than thirty seconds – while the physical letter sits in the sorting machine in Europe – to transcribe the information they see, and match it to a specific location in a geographic database. Should the Vietnamese worker not be able to code and match the address within thirty seconds, the sorting machine spits the letter out into an 'unsorted' pile, requiring a much more expensive European worker to manually sort the item. Therefore, even a small disruption on the Vietnamese side would massively slow down operations and increase costs on the European side.

Why, then, do we not see more strikes and withdrawals of labour occurring? The first part of this chapter discussed the many barriers and risks that workers face when organising. But there is also something else at play: the very opacity of these production networks. The reason why we refer to a 'European postal service' instead of revealing the location of the postal service is due to a strict non-disclosure agreement they enforce with their data provider in Vietnam. This is common the world over. Client firms do not want their suppliers to reveal anything about their work – even to their own workers. In interview after interview, we learned that workers typically had very little idea of who their client was or how the workers' activity fit into their business. The general absence of information about how supply chains work renders the possibility of targeted strikes to shut down production networks extremely challenging.

This is where the second strategy of sounding the alarm comes in. In situations where workers cannot 'block the flow', they may still have the ability to harm the reputation of their employer. Many companies invest enormous resources in constructing an image of themselves as progressive, ethical, and inclusive. If you walk into a Starbucks café anywhere in the world there is a strong likelihood that you will see a photograph of a worker in a field of coffee plants. They will likely be smiling amicably, carrying a small artisanal basket of coffee beans and leisurely strolling through a field. The message is clear: 'buying our product improves these workers' lives'. Certification schemes such as B-Corp and Fairtrade also help companies tell this story.

Recognising this, some trade unions have launched major public campaigns to present counter-narratives. There is a truly colossal range of tactics that unions deploy in this space, but what unites them all is an attempt to tarnish companies' reputations for consumers. In the UK, for instance, the IWGB union (a grassroots independent union formed in 2012) has attempted to convince the online grocery delivery firm Ocado to in-house all delivery drivers (rather than use third-party fleets) and meet demands for fair pay and just working conditions by encouraging consumers to use the #ShameOnOcado hashtag.[3] In 2023, a coalition of trade unions from Bangladesh, Cambodia, Sri Lanka and the US published an open letter to Adidas describing harms that workers face when working for their subcontractors. Addressing Adidas directly in the letter, rather than their employers, the unions note, 'We want to negotiate directly with you, the decision-maker at the top of the supply chain, and are open to any suggestions you may have on this proposed agreement.'[4] In response, Adidas claimed that while the subcontractor had previously made products for Adidas this was not the case at the time of the strike or after.[5]

One limitation with this approach is that these sorts of public 'name and shame' strategies are typically only effective against high-visibility companies. Companies like Meta and Google care deeply about their brand name and image, but some outsourcing firms further along the network may have less of a reputation to maintain. A large international coffee chain is naturally more concerned about what consumers think about working conditions than the companies that run plantations in Kenya, Indonesia and Brazil.

Here, unions of data workers face a critical disadvantage. Data work firms are rarely consumer-facing, as they are not running plat-forms, creating autonomous vehicles, or selling large language models. If unions, like the ACMU, are to leverage reputation as a tool to push for better working conditions, they must find ways of connecting conditions at their firms with the household-name companies that ultimately use their work.

Data workers therefore face an uphill battle if they are to secure concessions through negotiation with their employers. We have seen they have limited bargaining power, struggle to take advantage of strategic bottlenecks, and often lack the information required to publicise their stories through campaigns. It is therefore crucial to consider how alliances can be built and information shared across different groups of workers in AI production networks.

Tech-Worker Organising in the AI Production Network

Annotators are linked, in the same global production networks, with a certain segment of highly paid white-collar tech workers who occupy a distinct niche in the employment landscape. Until recently, trade unions have struggled to make significant inroads in the tech industry. In the United States, only about one in a hundred tech workers is a member of a union, and it is unlikely that rates are massively higher elsewhere.[6] By virtue of having relatively scarce and in-demand skills, most software engineers possess a large amount of bargaining power in the labour market.[7] As a result, they have typically been able to secure high pay and decent conditions simply because companies know they would otherwise lose these workers to competitors.[8] Salaries are especially high for engineers and developers specialising in AI. According to Business Insider, a typical AI engineer at Google makes approximately $254,701 a year.[9] At Microsoft, principal engineers can make upwards of $400,000; at Meta, the median overall compensation for AI engineers is $360,477; and at Apple even the lowest level engineers can make $200,000 to $250,000 a year with compensation. However, more recently, tech workers have begun to organise. Why? And what can we learn from those efforts?

In his summary of the tech worker movement, writer and tech worker Ben Tarnoff emphasises the election of Donald Trump as a critical catalyst that would ultimately lay the foundation for mass action.[10] As Trump made promises to his voting base that reeked of fascism, such as a pledge to build a digital registry of all Muslims in the United States, tech workers became painfully aware of the critical role they might be asked to play under a Trump administration. As a result, tech workers from around the US began to organise through a group called Tech Solidarity. Within a few weeks of the 2016 Presidential election, the group had already created 'the Never Again pledge' – a commitment by workers at firms such as Amazon, Apple, Google and Microsoft to oppose and not work on any databases for the US government that would target people by race, religion or national origin.[11] The name 'Never Again' is a reference to IBM's role in using its technology to assist the Nazi government during the Holocaust.[12] Despite the risk of losing their jobs by refusing to work on tasks, almost 3,000 workers signed the pledge within a week.[13]

This early moment in collective organisation in the tech sector was significant because it showed the impacts that such workers can have on factors that stretch beyond their immediate working conditions or pay. At first, out of all of the large tech firms, only Twitter agreed to oppose the incoming administration's plans to build a registry of Muslims.[14] But by the time the pledge had reached a few thousand signatories, a host of large companies such as Apple, Google, Microsoft, Uber, and even the company that had inspired the action's name – IBM – had made public commitments to never work with the government on creating such a tool. Tech workers had seen that they occupy a hugely important node in some of the world's most important production networks; and by organising around specific demands, had demonstrated they could strategically use their position in that network to create impacts that stretch far beyond their own working conditions.

In the years that followed, much tech worker organising has followed this formula. An organisation that has been central to these efforts is the Tech Workers Coalition (TWC). The group was founded in 2014 and grew rapidly – using forums and messaging tools to coordinate geographically dispersed members. Its explicit goal is to create a counterweight to the business interests of Silicon Valley, and to bring the labour movement into parts of the industry that traditionally have seen very little worker organising. In a 2018 presentation at a conference on worker resistance, one of its representatives described the organisation as follows:

We're mostly made up of people in various white-collar occupations in the industry: programmers, engineers, product managers, and so forth. But it's important to note that we really want to help organise the entire industry, across all occupations and stratas: everybody from cafeteria workers, to customer service reps, to data scientists. In fact, TWC originally started as a group whose main purpose was to help unionisation campaigns among service workers, and to enlist the support of the skilled technical workers at various sites. But since then, our ambitions have grown, especially as the experience of being in solidarity with service workers has led to more of us thinking of ourselves as workers as well, as part of the same struggle. So with regards to labour organising in and against platform capitalism, we're very excited and enthusiastic about considering the possibilities for leveraging the strategic position of skilled technical workers in the tech industry, in conjunction with the ongoing movements of what we could call 'platform workers'. In other words, we'd like to think seriously about the potential to build a class alliance between the workers that build platforms and the workers that use – or are used by – platforms.[15]

By the early 2020s, there had been a blossoming of worker organising, including a series of new worker-led organisations. In the US alone, there are now at least thirty-seven unions for tech workers, including the Amazon Labor Union (ALU) and the Alphabet Workers Union (AWU).[16] In June 2022, Microsoft pledged not to stand in the way of its workers organising and has been engaging in collective bargaining negotiations with the Communication Workers of America.[17] There have also been numerous attempts at workers organising through more established and older unions: for instance, contract workers at Google organising through the United Steelworkers Union and employees at Kickstarter organising through the Office and Professional Employees International Union.[18] By late 2023, the Collective Action in Tech (CAT) project (an initiative to document the diversity of collective action in the sector) had documented a total of 542 worker actions across the tech industry.[19]

The Tech Worker Coalition identified three broad issues around which tech workers have organised: standard workplace issues such as overwork, pay and stress; issues of diversity, sexism and racism in tech companies; and finally, ethical and political issues concerning what projects tech companies work on and their political stance on broader issues.[20] First, with workplace issues, overwork is especially prevalent for highly skilled workers like engineers and data scientists. Many tech firms encourage this overwork by providing employees with free or heavily subsidised food to keep them at their desks for as long as possible. For example, Chinese tech workers began organising against the gruelling '996 system' (9 a.m. to 9 p.m., 6 days per week), which had become an official work schedule of many mainland Chinese companies. But in early 2019, the 996.ICU campaign, named after the suggestion that following this work schedule would send anyone to the ICU, began to take form.[21] In response to 996.ICU, a group of Microsoft employees penned an open letter in support of the Chinese movement.[22] The letter was crafted to both demand that Microsoft resist any attempts at censorship (Microsoft owns the GitHub platform on which workers organised), and to express international solidarity – with signatories hailing from around the world. The letter concluded with the following statement: 'history tells us that multinational companies will pit workers against each other in a race to the bottom as they outsource jobs and take advantage of weak labor standards in the pursuit of profit. We have to come together across national boundaries to ensure just working conditions for everyone around the globe.'[23]

There are also workplace issues faced by outsourced workers at tech companies. Across the tech sector, huge numbers of blue- and white-collar workers are technically employed by third-party agencies. Often lacking the pay, benefits and employment security of their in-house colleagues, these workers are a second-tier workforce that tech firms use to fill the gaps in their roster or play essential but lowly remunerated support roles. In 2019, the majority of the 220,000 people working for Google across the world fell into this second camp.[24] This striation of the workforce draws on a long history of employers reinforcing various distinctions among employees in order to cut costs and increase control.[25] It presents organisers with serious challenges, but can also offer unconventional opportunities. In response to poor conditions, outsourced workers and their in-house allies have coordinated campaigns to secure improvements. For instance, in 2019, when Google reduced the contracts of contract workers who helped to build the 'personality' of Google Assistant, over 900 workers signed a petition demanding the conversion of contract workers to full-time direct employees and the extension of associated benefits. This agitation on the part of in-house workers resulted in the announcement by Google that agencies will have to pay contract workers based in the US at least $15 per hour and offer them comprehensive health insurance.[26] This action has shown how pressure can be put on lead firms in production networks to improve conditions for workers employed by other firms in the same network. If conditions are to improve for the millions of data workers around the world, this tactic will need to be deployed not merely in the US, but across international borders too.

A second type of issue, as identified by the TWC, around which workers have mobilised is diversity and inclusivity issues, including sexism, racism and sexual harassment claims. In 2018, over 20,000 Google workers in fifty cities around the world staged a walkout over a cover-up of sexual harassment claims.[27] Writing in *The Cut*, seven of the organisers outlined five key demands: an end to forced arbitration; a commitment to end pay and opportunity inequity; a publicly disclosed sexual harassment transparency report; a clear, uniform, globally inclusive process for reporting sexual misconduct; the promotion of the Chief Diversity Officer to answer directly to the CEO and for the appointment of an Employee Representative to the board.[28] Although the scale of mobilisation was impressive, almost none of the demands of workers were ultimately met. A year after the walkout, the *Los Angeles Times* published an analysis showing that only (part of) one of the five demands was met by Google.[29]

Finally, workers have organised around larger ethical and political issues. One of the most high-profile examples of this is the action taken by Google employees against their company's involvement in a US Military project to analyse drone footage using machine learning, dubbed 'Project Maven'. Many workers at the company felt that the project clearly fell afoul of the 'Don't be evil' motto that Google had promoted since the early 2000s.[30] What began as expressions of concern on an internal company social media platform quickly evolved into an open letter, signed by 5,000 workers, addressed to Alphabet CEO Sundar Pichai demanding that Google pull out of the project.[31] This campaign against Maven was the culmination of earlier efforts around the Never Again pledge. By getting their employer to stop working for the Pentagon, workers saw they had meaningful collective power, and that through organising they could effectively wield it. As one of the leaders of the campaign put it, 'Ultimately, the Project Maven campaign wasn't just about whether Google should build this one tool for the military. It was about using our power as workers to ensure that technology is built for social benefit and not just for profit.'[32] The workers were ultimately successful in pressuring Google not to renew the Maven contract.

As with the Never Again pledge that came before it, the anti-Maven action galvanised workers elsewhere in the industry to make changes at their own workplaces. Workers at Microsoft followed suit, calling for the company not to bid on the Joint Enterprise Defense

Infrastructure (JEDI) in 2018 and for the cancellation of a 'contract to provide augmented reality headsets to the US military' in 2019.[33] 'We did not sign up to develop weapons, and we demand a say in how our work is used', read part of the letter.[34] Neither was successful, with Microsoft ultimately winning JEDI.[35] Still, it has not stopped further demonstrations of resistance. In 2022, Amazon and Alphabet workers came together with community organisations to protest Project Nimbus – a $1.2 billion contract with the Israeli government for the provision of cloud computing services that was announced in May 2021.[36] The campaign, #NoTechForApartheid, has orchestrated protests at several Google HQs as well as events such the Google Cloud Next Conference and the Amazon Web Services (AWS) Summit.[37]

Like data workers, high-skilled tech workers occupy both topological and temporal bottlenecks in the AI production network. If they choose to withdraw their labour power en masse, they will cause enormous economic disruption to their employers and the broader AI production networks. Yet, interestingly, despite the myriad attempts to organise workers, there were very few attempts to *block the flow* through actual strikes and work stoppages. Much of it instead manifested through *sounding the alarm* in the form of petitions and public statements.[38] Firms worry both about the bad publicity generated when their workers organise, and the real risk that worker organisation might escalate and lead to broader and more painful industrial action.[39] By presenting firms with a combination of these risks, tangible victories have been achieved by organised workers.

Unlike data workers, high-skilled tech workers possess two key advantages. While data workers are hindered by extreme levels of opacity within the network, high-skilled tech workers are typically more familiar with how their work is being used, affording them the ability to either disrupt or effectively threaten to disrupt key processes. Second, while data workers are competing in a planetary labour market against a reserve army of labour, most tech workers have in-demand skills. A company might consider firing a few thousand data annotators in one place and hiring a few thousand more in another, but no large company would contemplate doing the same thing for their teams of machine learning engineers. As we have seen from some of their successes, this basic fact affords these workers greater power.

A Planetary Workers' Movement

How can highly paid tech workers join together with data annotators and content moderators to ensure that the most vulnerable workers in the system are afforded decent pay and decent jobs? Just as their employers use the planetary system against workers, workers will have to find ways of turning the planetary network to their own advantage. To do this, they will need to think about their jobs as not just tethered to a particular place, but as part of complex, globe-spanning production networks; and they will have to use that framing to think, strategise and act transnationally. But which organisational forms of transnational solidarity will be most effective in achieving the demands of workers, particularly those most vulnerable in AI production networks? There are three ways that workers can organise transnationally: acting in solidarity, transnational alliances and trans-national unions. What can each of these do for workers?

The first example involves workers acting in solidarity with groups of workers in another part of the world. Over a hundred years ago, for instance, British dockworkers at London's East India Dock refused to load a ship with munitions which were being sent by the British government to the Polish government, for the Kiev Offensive to repress Soviet Russia following the revolution. Since then, other groups of workers who work at strategic bottlenecks have found ways of expressing their solidarity. Following the Pinochet coup of 1973 in Chile, for example, dockworkers in Liverpool refused to unload any food from Chile, to put pressure on the dictator's economy. In addition, Rolls-Royce aerospace engineers in East Kilbride refused to repair jet engines for the Chilean air force (as immortalised in the film *Nae Pasaran*). Elsewhere, before the 2003 invasion of Iraq by American and British forces, railway workers in Scotland refused to drive trains to a weapons depot. The drivers that took action did not face disciplinary action, in part because of the strength of their trade union. And, during the 2023 invasion of Gaza by Israeli forces, multiple arms factories across the UK that provide arms to Israel were shut down by trade unionists who formed a mass picket in front of the factory gate.

While these campaigns have the potential to be highly effective, they ultimately rely on the compassion and altruism of people in one place to want to help those in another. They can certainly deliver concrete wins to groups of workers who otherwise might lack the leverage to push for change themselves. However, campaigns led and coordinated from the Global North to deliver results for workers in the Global South will always remain one-sided. Compassion and altruism can be powerful forces when marshalled, but they will never structurally change the balance of power within an entire production network.

A second way of organising across a production network is through transnational alliances in which there is more of a reciprocal relationship between members. These transnational alliances can include unions, worker associations, civil society and worker advocates. A key role that these transnational organisations can serve is to negotiate with, and put demands to, transnational firms in a way that stretches beyond the demands of any particular local group of workers. A transnational alliance can be any coalition or confederation of groups in which local organisations remain distinct, but they coordinate strategy on an international level.

Transnational alliances between groups of workers in different countries have also existed for some time. This is, in part, because they have grown into the gaps left by the withdrawal of the state in regulating labour in low- and middle-income countries. Governments in such countries, inspired by neoliberal ideology and disciplined by the stringent terms of financial support from the International Monetary Fund, have instead often seen labour as a source of comparative advantage. Regimes that repress and sometimes even permit violence against trade unions seeking to lobby for better working conditions do not create conducive environments for local trade unions to unilaterally extract meaningful concessions within global production networks. As such, transnational coalitions have begun to develop global framework agreements with transnational companies that seek to impose minimum standards of decent work across entire global production networks.

The Clean Clothes Campaign (CCC), for example, is a global network of over 235 organisations in forty-five countries that seeks to empower and improve the conditions of workers within global garment supply chains.[40] In the aftermath of the tragic Rana Plaza factory collapse in Bangladesh in 2013 that killed 1,134 workers in five garment factories, the CCC lobbied for the Accord, which was a legally binding agreement on health and safety in the garment industry.[41] The Accord includes a number of important safety protections, safety inspections, and also affords workers the right to refuse to do unsafe work. Importantly, if inspections cause any supplier factories to be shut down for remediation and maintenance, upstream fashion brands are contractually responsible to compensate workers during the period of closure. The second iteration of the Accord took effect in 2021.[42] As of 2023, there were over 200 signatories. This is the kind of agreement that individual local unions alone would struggle to extract from a large multinational fashion brand.

ExChains is an example of a transnational initiative that has sought to build local power as the foundation to cross-border solidarity. Formed in 2002 by trade unionists in affiliation with ver.di (Germany's second largest trade union), ExChains is a network that connects factory workers from garment-producing countries across Asia with retail workers in Germany.[43] Unlike NGO-led labour initiatives such as the Clean Clothes Campaign, ExChains is 'grassroots' – workers and rank and file trade unionists shape the agenda and strategy of the organisation.[44] Shared interests between the various international stakeholders are central to the formation of worker power at the local level. This strategy lays the foundations for future international worker solidarity by allowing workers to understand the conditions of work in different parts of the production network. International solidarity between workers can never be built without developing those understandings.

Ultimately, though, ExChains' emphasis on building local power reflects the fact that there are countless examples of firms that do not see workers as a partner to engage in dialogue with, or an actor with whom they might co-determine the nature of jobs. Rather, they are viewed simply as a commodity to be bought and directed; a resource from which as much value as possible needs to be extracted. Some of these firms may be vulnerable to the types of pressure that can be exerted through solidarity campaigns or via transnational alliances. But, because of the opaque nature of production networks, many will not be. For those firms, striking is the only tactic they respond to.

The third way in which pressure can be brought to bear by workers around the world is through a transnational union. This would involve workers from different parts of the world within a single production network building collective power by agreeing to act as one organisational entity. Through this approach, they would be able to do something that neither of the two previous strategies could quite achieve: directly coordinate workers in selective strikes. It is important to note that while the idea may seem far-fetched, it takes inspiration from existing examples of powerful workers organising in large and complex networks.

A selective strike is a strategy in which rather than all workers walking out of the job, only a few do. The workers who withdraw their labour power during a selective strike are workers who need to occupy key bottlenecks in their company's production process or service provision. Examples of selective strikes from a single country can help exemplify how this strategy might work on an international scale. The approach was first pioneered in the wave of industrial action that began in Europe after 1968. Strategic workshops within factories would shut down for short periods in order to cripple the production of the whole plant. It was formalised as a tactic by the Association of Flight Attendants in the early 1990s, when they developed a strategy that they dubbed CHAOS, or 'Creating Havoc Around Our System'. CHAOS involved only a few flight attendants walking off the job just as passengers began boarding a flight. Because airlines couldn't predict which flights would be hit, they eventually caved into the demands of the flight attendants.[45] More recently, this strategy of using leverage has been put into practice by the United Auto Workers union (UAW) in its 2023 campaign to win four-day workweeks, pay raises, and a number of other benefits for its 145,000 workers who produce about half of the cars made in the United States. The UAW would selectively shut down the supplier of a key part or shut down a factory producing a model of car known to be in short supply. Similarly, in Germany, the ver.di trade union – which represents almost two million workers – is famously able to rely on workers in some of the most potentially disruptive positions in society to strike to achieve better conditions for millions of other workers. In 2023, a selective strike by workers who attend to garbage collection and sorting (and just think what your city would be like after a few weeks if the garbage collection system stops working!) was used to demand a 10.5 per cent pay rise for millions of other workers.[46] All of these examples of selective strikes illustrate how unions can shut down an entire production network, or cause enormous economic and social disruption, with the fewest possible workers: meaning that strike funds are preserved and risks to workers are minimised while still maximising leverage with the company that they seek to extract concessions from.[47]

The problem with all of these examples of selective strikes is that they occur within a single country. A fact that is not particularly useful for workers like Paul and his colleagues. While there are a lot of workers in Paul's position, the work they do almost immediately leaves the country as soon as they submit it. Only a relatively small part of the production network is therefore local. If the likes of Amazon and Google can coordinate pieces of their production network on a planetary scale, then workers too will need to build power at an equal scale.

What then are the possibilities for other trade unions to do what ver.di and the UAW are doing? How could you develop a CHAOS strategy for the AI sector? And build a union across an entire production network? To do so would require a single union that would not just share information from around a global network, but which could count on a transnational membership of workers and coordinate them through selective strikes that strategically targeted vulnerabilities in the network. Such a strategy has never been achieved. This is, in part, because existing parochial unions would naturally be resistant to losing due-paying members to a larger organisation, potentially based in another country. It is also, in part, because a transnational union would struggle to operate outside of the existing national-scale legal frameworks that currently regulate trade unions. In the UK, for instance, business-friendly trade union laws restrict trade unions from engaging in disputes against any entity other than their employer (a big problem for unions if they are trying to engage in effective select-ive strikes at the scale of a global production network).

So while the idea of a transnational union is certainly far-fetched, neither of the above-mentioned issues is quite as insurmountable as they initially seem. A transnational union for AI workers is moving into a space characterised by relatively few existing organisations. Many large trade unions have massively shrunk during periods of deindustrialisation, and many never organised workers in the data work industries in the first place. A transnational union would still likely be able to negotiate the myriad national-scale regulations that govern how it can organise its workers. It, after all, would not be based in the cloud. Each of its chapters would be firmly located in places where it is indeed possible to organise workers. The strategy would be fraught with difficulties. But the path to worker power has never been easy.

We have discussed three broad strategies that can be deployed to improve the jobs of data workers. Acting in solidarity is a way in which power can be non-reciprocally used by one group of workers to help another. It allows the strong to help the weak. Transnational organisations seek to offer a counterweight to transnational companies at the global rather than local scale. While there are many examples of large unions organising across sectors or production networks, there is still much to be done to build unified unions that might organise workers at the planetary scale. If workers are going to win in a global network, they are going to have to work together globally.

Data workers cannot rely on altruistic behaviour or solidarity from other workers in faraway places. Altruism is not a structural solution to the exploitation that is baked into digital capitalism. But at the same time, they should not ignore or disregard it. There are many examples of transnational solidarities, both within the tech sector and outside of it, that have made a meaningful difference to the lives of workers. Those solidarities should, and must, be harnessed. But much more is needed. Workers the world over will need to explore ways of organising at a planetary scale: whether through single campaigns, transnational federations, or through a yet-to-be-built transnational union.

Workers are experimenting with a diverse spectrum of organising efforts: from media campaigns, to petitions, to strikes, to ambitious efforts to organise across supply chains. Pressure is being exerted on policy-makers and through supply chains. Workers are appealing to the hearts of consumers and decision-makers. They are striking fear into the hearts of managers. And they are doing all of this by organising. It is ultimately their experimentation that will determine the path towards building power in the sector. While we can offer our informed speculations here, the future is in their hands. It is only through collective organisation and action that their interests can be served. If workers don't stand up for themselves, then nobody else will.

8

Rewiring the Machine

Some of our most striking images of AI are found not in chatbots, search engines or autofill software, but in science fiction. One particularly evocative description is in Scottish author Iain M. Banks's Culture series of novels. Banks writes about a utopian, post-scarcity society many years in the future. The reason for this society's great wealth and abundance is their creation of advanced super-intelligent AIs that perform most of the work. The absence of the drudgery of work helps the humans create a utopia: there's no money, no wage labour, and nobody wants for anything. People's lives are full of luxury and hedonism; it is everything for everyone. But most of all, they are free.

Science fiction can play an important role in inspiring us to think about different possible futures of technology, provoking us to explore new ideas and ways of being that might not seem possible or even imaginable within our current society. The philosopher Ernst Bloch wrote, 'The most tragic form of loss isn't the loss of security; it's the loss of the capacity to imagine that things could be different.'[1] Sci-fi allows us to challenge existing social norms and power structures through concrete visions of how the world could be organised along different lines.

Today, we live in a very different world from the one Banks imagined. Machines create art, compose music and write poetry, while countless humans are forced to work like robots, toiling in monotonous low-paid jobs just to make such remarkable machines possible. AI is often viewed as an inevitable force that we must adapt to rather than a tool that we can consciously shape through our own actions. We live in a time of technological determinism in which tech entrepreneurs loudly proclaim how their wondrous inventions will change our lives for the better. But the millions enlisted to build, service and repair these machines are part of a hidden army of workers kept behind the curtain by companies interested in maintaining the fiction of smart and autonomous machines. How do we create a world in which machines serve humans rather than one in which humans are the servants of machines?

We do not think such a world can be brought about through any single act or with one particular group working in isolation. In this book, we have seen how power is centralised within the production networks of AI. Workers, from data annotators and content moderators to warehouse operators, artists and machine learning engineers, all occupy very different nodes in these networks. But they all share one thing in common: they all have a negligible amount of control over the structural conditions of their work. This includes how their work is organised, the conditions that govern their jobs, the amount of value extracted from their labour, and the impacts of their work on the wider world. They are all plugged into a system that co-ordinates a vast planetary-scale division of labour; a system that is able to extract not because it renders labour abstract, and treats everyone everywhere as an interchangeable cog in the machine, but because it carefully segments the world's workers.

To each of the workers introduced in this book, the production networks of AI are a black box. There is no map to make sense of this territory, and the tendrils of the network stretch far beyond the vision of even the most connected actors in the system. While there are lead firms who coordinate large amounts of activity within each network, the system of production is much bigger than even them. The risks and harms produced by this system are therefore outside of the scope of any single actor to rectify or improve. Instead, if we are to think about production networks of AI that are more human, just, decent and fair, we should engage in a range of overlapping strategies. These strategies will need to be enacted in different places, at different scales, and by different actors. But, together, they stand a chance of changing not just a part of the system, but the system itself.

In this chapter, we outline five concrete steps to guide us in the right direction towards a world in which all AI workers are afforded dignity and respect. These steps focus on the economic conditions of workers within AI production networks and do not necessarily extend to changing AI models themselves or rethinking the technical aspects of AI systems. We leave this task to others better placed to make such interventions. The steps we propose are: building the collective power of worker organisations; civil society organising to hold companies accountable; enacting stricter regulations on AI companies; establishing forms of worker governance and ownership over enterprises; and challenging the injustices of the broader system within which these companies operate. Each of these steps would go some way to creating fairer conditions for workers in these systems and can be pursued concurrently. Progress towards one of the goals will often assist with achieving the others. For example, more stringent regulations for AI companies on how they outsource tasks to contractors in the Global South would empower workers and strengthen their position in negotiations with employers. If these steps are taken together, we can create a society in which AI works for all of us rather than humans feeding the machine.

Step One: Build Worker Power

At the basis of the problems facing many of the workers we have encountered in this book lies a lack of power to make meaningful decisions over institutions that have a significant impact on their lives. Workers find themselves as small parts of complex and interconnected systems that stretch across the globe. They face the immediate disciplinary power of their bosses at work, but are also subject to larger forces that constrain them through pressure exerted by supply chains and the production networks of AI. Many data annotators work under strict regimes of workplace control, but even supervisors within these digital factories feel compelled to enforce substandard working conditions to keep their firm's contracts from going to its competitors.

At an individual level, a worker has their own job and, depending on the workplace, may have a degree of autonomy over their work-flow, but all of the big decisions about how to invest scarce resources and the fundamental direction of the company are made by investors and management. The degree of disempowerment is relative, of course, but everyone from the annotator to the machine learning engineer lacks a right to formally participate in decision-making at work and confronts a significant power difference between themselves and their employer. On their own, even the most productive and well-paid engineer has little to bargain with if they find themselves at odds with their boss. Workers are structurally disempowered and experience a lack of control at work because, as individuals, they are replaceable, and what they bring to the table – their skills, training and labour – can be found elsewhere.

The experience of modern workers is not unique. Social groups throughout history have struggled against being exploited and governed by a ruling class. Most societies have relied on some form of economic system in which a minority benefits from the labour of the majority – whether it's through peasants tilling the land for an aristocracy, assembly workers on the production line for factory owners or data annotators helping train AI models for tech companies. Historically, we have called this *the problem of oligarchy* – a small group of powerful elites naturalise their rule over the majority and appropriate value from their labour. Once ruling groups achieve power, they consolidate their position by building powerful institutions and systems to defend it. We can think of many such systems across different historical periods – of men over women, those racialised as white over people of colour and European colonisers over non-European peoples, to name a few.

Dominant social groups maintain these systems through force, by dividing those they dominate and convincing them that the system operates in their best interests.

When social movements have succeeded in enacting change they have done so primarily by organising the majority to act against the wealthy and powerful elite. Democracy, as it emerged in ancient Athens and other smaller city-states in the region, was a class-based movement of poorer Athenian citizens against the dangers of the corrupt rule of a few wealthy Athenian families.[2] Decolonisation movements of the twentieth century in former colonial territories were often driven by powerful union movements of workers coming together to liberate themselves from their colonial oppressors.[3] The civil rights movement in the mid-century United States made considerable gains after mass mobilisation campaigns against a white ruling minority that had legalised segregation and disenfranchisement.[4] Historically, social movements that brought home the goods did so not by appealing to the kinder nature of their oppressors, but by building their own power and forcing change through political struggle. As the Frederick Douglass quote goes, 'Power concedes nothing without a demand. It never did and it never will.'[5]

This is what we call the *unequal power* view of social change: the only way to secure lasting social change is through oppressed social groups building their collective power in movements and institutions. This view is in opposition to those reformers who have argued that autocratic rulers could be voluntarily persuaded to make reforms by appealing to principles of fairness and justice without any recourse to coercion or struggle. Radical campaigners as diverse as Nelson Mandela, Rosa Luxemburg, Martin Luther King Jr. and Angela Davis have all believed that genuine social transformation would only occur following a fundamental shift in the balance of power between social groups.

Of the seven characters introduced in this book, all but the investor share a common interest in changing the system so that value and power is more equally distributed between workers. Although they work under radically different conditions due to how much power they have in the network and how their job and country of birth affect their experience of work, all of these workers share a common thread of being pulled into the extraction machine to produce AI systems worth far more than any of them are compensated for. They all lack the power to transform how the broader network operates or to make decisions about their own work and role in the production of AI. This points to a fundamental antagonism between a class of investors and bosses who govern this system and those who must obey their commands. In our global capitalist system, companies are incentivised to make AI that increase their profits. This factor is the single most important determinant of how AI is developed and the working conditions of those who are part of the system.

The first step for workers in the extraction machine to develop their collective power is to join a union to secure better pay and conditions, protect their rights and provide a counterbalance against their bosses' arbitrary power.[6] When these workers face large companies on their own, their power is negligible. It is only through joining forces with others as part of collective campaigns that workers can defend their shared interests and achieve strategic goals. Unions harness the collective power of workers to negotiate collective bargains on working conditions. Yet they also exercise an important political role, asserting the equality and dignity of workers and challenging the power that the wealthy exercise over others in the absence of countervailing forces.[7] At their best, unions defend the principles of freedom and equality and strive for a society in which we all feel empowered to assert our equality and defend our interests. They are the most important weapon workers have in the struggle for fairer work.

Alongside unions, another mechanism for enhancing the power of workers is to enable workers' elected representatives to participate in decision-making at the level of the firm in what is called co-determination governance. This is a practice where workers have a right to vote in their own representatives to a board of directors at their company, providing them with a voice in company decisions. This could involve roughly half of a board of directors consisting of representatives of management and the other half consisting of workers. Such arrangements are most well-known from German

Mitbestimmung (co-determination) schemes introduced in 1918 after German workers mobilised for greater democratic control in the workplace.[8] In some models of co-determination, workers also form 'works councils', with worker representatives involved not only in director-level decisions but in more day-to-day discussions about how the firm operates. These schemes are not an answer to all the problems workers face, but they have been found to increase workers' wages, enhance job security and strengthen the power of workers in the firm without harming its competitiveness.[9] More recently, Belgian sociologist Isabelle Ferreras has proposed a bicameral system of corporate governance that would enable workers to share in the governance of a firm and grant them equal rights to capital investors.[10] Since investors and workers both contribute important resources to the firm, they should each have an operational representative body with veto power, with major decisions for the firm requiring a majority of votes in each of the bodies.

In addition to changing the balance of power in how AI firms are governed, we could also look at democratising the value produced by firms by instituting an employee ownership scheme that would grant workers partial ownership over their firms. According to this idea, private companies employing a certain number of people would have to set up funds that would offer their workers a significant financial stake in the company. An early attempt at this was put forward by the Swedish economist Rudolf Meidner in the 1970s.[11] Under the 'Meidner Plan', a gradual transfer of ownership of capital from wealthy shareholders to workers was envisaged through a system of wage-earner funds. Despite its initial support, the plan faced strong opposition and was never fully implemented. Inclusive Ownership Funds were also suggested in a report by the UK-based Institute for Public Policy Research and supported by the British Labour Party during its 2019 election campaign.[12] Then Shadow Chancellor John McDonnell suggested that 1 per cent of shares of a company would be transferred to an employee ownership fund each year for a period of ten years until workers had a 10 per cent stake in the company.[13] The idea behind this is that instead of just rewarding workers with wages, they will have a financial stake and an ownership role in the company. This would assist in ensuring a fairer distribution of economic power and the ability of workers to have greater decision-making rights at their workplaces.

Of course, there are many complexities with plans for greater union-isation and building workers' power through employee-ownership schemes. Internal tensions can arise between rank-and-file union members and the union's leadership structures, in addition to diffi-culties organising across sectors and across national lines. We have already seen in Chapter 7 the importance of fostering a truly trans-national workers' struggle that links together blue- and white-collar workers throughout the global production networks of AI. Building worker power could take a variety of forms in addition to the exam-ples we have provided here. If any worker wishes to wrest back the ability to shape their jobs, be they an annotator, an operator, an artist, a technician or an engineer, and thereby the ability to shape the future, the first step is building collective power.

Step Two: Hold Companies Accountable

In addition to building workers' power within workplaces, there are important openings for consumer and social movement-led pressure to be exerted on companies to improve pay and conditions for all workers throughout the supply chain. This form of action is very different from collective worker action because it relies primarily on outside forces using the reputation of the company as the key point of leverage in making their demands. As we saw in the previous chapter, such 'name and shame' campaigns have proved effective in holding companies accountable for their actions and helped raise the standards of their business practices. Nobody wants negative PR if they can avoid it. Companies will go to considerable lengths to ensure they stay out of the spotlight. There are a few actors in these AI production networks that have the power to shape and influence working conditions across the entire network, and these companies can be vulnerable to targeted campaigns that encourage them to use their power in less harmful, and possibly even responsible, ways.

While there are numerous examples of civil society-led campaigns, we offer one here that is deeply connected to several stories in this book and within which we have been personally involved. This example is of 'Fairwork' – an action-research project that holds companies accountable for how platform-based technologies are used in workplaces. Fairwork has designed a set of ten 'fair work principles' as objective benchmarks that we evaluate each company against: giving each company a score between 0 and 10.[14] A zero score means that a company cannot evidence that they adhere to even a single standard of fair work; whereas a ten indicates that each minimum standard has been met. That score is then used as a lever to encourage firms to improve pay and conditions in subsequent negotiations with them over how their score could be improved. It does this by not allowing companies to opt out of a rating, and lets them know that they will be scored on nationwide comparative league tables. The project started in South Africa and India in 2018, with funding from the German Ministry for Economic Cooperation and Development to evaluate digital labour platforms.[15] Since then, the project has expanded to thirty-nine countries, scored companies 618 times, and – in doing so – has pressured companies to make 299 improvements to the jobs of workers.[16]

In 2023, the Fairwork project evaluated an AI data annotation company called Sama. By integrating data annotators in East Africa into the AI production networks of large Fortune 50 companies in the Global North, the company claims to be empowering tens of thousands of people. 'Give Work, Not Aid', the website states in bold letters. If you only knew the company from its website, you would indeed have a vision of a firm that is a force for good in the world. The website presents image after image of smiling workers in offices full of green plants, and statistic after statistic espousing the difference that the company has made: '65K+ Lives impacted since 2008', '25K Skill building trainings completed in 2022'.[17]

We wanted to apply the Fairwork methodology to Sama to see whether the lived reality of workers actually matched up to the company's bold claims about supporting 'ethical AI'. After meeting with dozens of workers in Kenya and Uganda over the course of a few weeks, we uncovered a company that fell far short of the basic standards enshrined in the Fairwork principles. Worker after worker spoke to us about a host of workplace issues, sometimes in very emotional conversations in which they were reliving harms they experienced, told us about being paid below the living wage, being forced to work unpaid overtime, being stuck on short-term precarious contracts, and a whole host of other issues.

We tallied up Sama's score based on this evidence and presented them with their Fairwork scorecard. It was a 0 out of 10. But, rather than making the score immediately public, we gave the company a couple of months to implement changes that would improve their score. For each one of the ten fairness thresholds, the company received a short wishlist, and, remarkably, within two months they had implemented twenty-four improvements to the jobs of their workforce. There were some important changes here that evidenced a genuine desire to improve conditions at the firm. These changes included guaranteeing at least the living wage to all workers, extending contracts (often moving workers from one month to twelve-month contracts), and eliminating unpaid overtime.[18] Those twenty-four changes brought Sama's score up to a 5/10.[19] This is a massive improvement over where the company started, but is only halfway to the minimum standards of decent work that a 10/10 represents.

The 299 cumulative improvements to the jobs of workers achieved by the Fairwork project have largely come about because the scorecard approach resonates with company managers. Those 299 improvements, whether they be guaranteeing wage floors, the provision of health insurance, or the implementation of anti-discrimination policies, are all changes that managers have chosen to make as a result of dialogue with Fairwork teams rather than their own workers. While it is patently ridiculous that companies would rather listen to a group of academics telling them what to improve rather than their staff, it illustrates just how powerful this sort of pressure can be.

Companies have made those 299 improvements to jobs because they understand that a poor reputation can be a significant vulnerability for them. A poor reputation, especially when that reputation can be boiled down into an easy-to-understand numerical score, can result in a loss of customers and thus a loss of business. In the age of Environmental Social Governance (ESG), corporate responsibility and so-called 'responsible capitalism', many companies, especially in the tech industry, go to great lengths to present themselves as benevolent, ethical and positive forces in the world. Now, whether or not you believe that 'responsible capitalism' is an oxymoron, what matters for the purposes of pressuring companies to improve conditions for workers is that the companies themselves believe they should be seen to act responsibly.

It is through this opening that we see hope for transforming significant parts of AI production networks. A number of lead firms already impose minimum standards on their suppliers. For instance, as part of their respective supplier codes of conduct, Google, Microsoft and Amazon all stipulate that anyone working for their suppliers cannot log more than sixty hours a week on the job. Google further specify that 'Employees must also be allowed at least one day off every seven days'.[20] Meta demands even higher standards for their contingent workforce: insisting that their suppliers provide their workers with at least fifteen paid days off per year, and a $4,000 child benefit option to facilitate paid parental leave. However, they only apply these stipulations to their contingent workforce in the United States. These standards are not imposed on their international suppliers.

Although these standards are depressing in their ambition (especially so, given that sixty-hour and a six-day work weeks have not been the norm in the Global North for a century or more), they illustrate that lead firms in AI production networks have already begun the process of regulating standards within the complex networks they coordinate. The fight to improve working conditions therefore does not have to convince companies that they can, and should, pressure suppliers to impose minimum standards. They already do that. The fight instead can focus on the details: raising wages and reducing working hours.

From these openings, much can be done. The example of Fairwork shows that the reputational sensitivities of companies can be leveraged to push for change. Workers are most vulnerable at upstream points in the network where they suffer the most direct harms and have the fewest direct mechanisms to lobby for change themselves. But strategies to hold lead firms accountable can benefit the jobs and lives of workers elsewhere due to the power these lead firms have over the entire network. Creative workers will seek to pressure lead firms to establish guardrails to protect their work from being stolen and uncredited, and from AI replacing artists altogether. Operators will seek to ensure that work is not unreasonably intensified. And engineers will seek to ensure that there are meaningful ethical guardrails in place for the products they develop.

Civil society can work together with the labour movement to map and link together the complex nodes in the network so that lead firms can be pressured to improve conditions for workers in faraway places. Few lead firms will want to be framed as unethical when faced with sustained pressure, but that sort of leverage will never work if strategies are not constantly developed to connect the dots between the different actors who play a role on the global map of AI.

Consumer pressure that is neither grounded in the embodied collective power of workers nor the institutionalised power of regulation will always be ephemeral. It has no inherent stickiness nor staying power. By itself, this strategy is therefore insufficient to rebalance power in the production network. But, as part of a broader suite of efforts to raise standards, improve wages and make the production of AI more fair, applying pressure on large firms can be an effective strategy for change.

Step Three: Government Intervention

Steps One and Two demonstrate how pressure from workers and consumers can be deployed to counteract the power of corporations in AI production networks. Both strategies can be effective, but they also have significant limitations. Without statutory regulation on the books, companies will always be tempted to push back against attempts to curtail their power.

The first issue for creating effective legislation involves thinking more carefully about jurisdiction. AI production networks are global, but laws have to be made in particular locations. As much as tech firms might like to think of themselves as located in the cloud, they are always based in real-world jurisdictions and subject to their laws. By considering the geography of particular production networks, we can assess the role laws in different jurisdictions might play in shaping outcomes. Three geographies of regulation are most relevant: *upstream regulation*, *downstream regulation* and *network-wide regulation*.

Upstream regulation – in the jurisdictions in which low-wage work is primarily carried out – is simultaneously the most straightforward and the most difficult to achieve. Jobs like data annotation and content moderation tend to be located in places where wages are low and regulation is lax. Companies that bring jobs into a community and are ready to leave if conditions for them deteriorate can actively deter serious attempts to regulate them. Some governments are also unwilling to apply stricter regulations for fear of these companies leaving.

With over a million workers in the business process outsourcing (BPO) sector, the Philippines is an interesting case of a country that is stuck between a need to improve the jobs of a huge number of workers and a country that knows that it cannot afford to kill an industry providing work to such a huge proportion of the population. In 2014, we had been in the sprawling metropolis of Manila for a few weeks, doing research on the country's enormous business process outsourcing industry. Towards the end of the trip, we managed to secure a meeting with a senior official at the country's Information and Communications Technology (ICT) Office whose mandate included fostering economic development in the country. A key question on our minds was what could be done to improve the quality of those jobs. But rather than discuss strategies of economic upgrading that would allow Filipino firms and workers to climb the value chain, we encountered an entirely different perspective. 'We're a country of one hundred million Filipinos, okay? . . . We're graduating more than half a million Filipinos every year,' he told us – indicating that the quantity rather than simply the quality of jobs mattered. 'Whatever they're doing right now, and they're good at it, let them stay there.' Low-wage work, in the eyes of this official, was an integral part of the Philippines' strategy to bring jobs and income into the country. Raising wages therefore was simply anathema to what the government was trying to achieve in nurturing the sector.

While the ICT Office is mandated to create ICT-related jobs (and has no mandate to focus on the welfare of workers in those jobs), others within government have nonetheless been trying to improve conditions for all BPO workers. Since 2009, there have been several attempts to bring in legislation to specifically 'protect the rights and promote the welfare' of workers engaged in the BPO sector. The BPO Workers Welfare and Protection Act, first filed during the 14th Congress, seeks to mandate employer adherence to the Labour Code, prohibit abusive or violent treatment and protect workers against understaffing and overloading. The proposed act demands that BPO companies provide all workers with access to information regarding their rights and benefits and the agreements between client and vendor as well as ensuring that workers are made employees after a training or probation period of a maximum of six months. Many versions of this bill have been introduced, but all remain pending at the time of writing.[21] The threat of destroying an industry that provides work to so many Filipinos looms large over any discussions about regulating the sector.

The economist Joan Robinson famously said, 'The misery of being exploited by capitalists is nothing compared to the misery of not being exploited at all.'[22] That statement rings true when seen in the light of the wicked choices faced by policy-makers in much of the Global South who feel that they have to choose between bad jobs and no jobs. The planetary labour market for annotation, moderation and other BPO jobs fuels a race to the bottom and ties the hands of policy-makers who seek to institute any truly transformative changes in the sector that might benefit workers.

For professions where there is a certain amount of geographic stickiness to their jobs, the story is different. For those jobs, offshoring is more challenging to contemplate. In Italy, while employment legis-lation is meant to provide only a basic regulatory framework, specific aspects are left to agreements between the social partners. However, governments have routinely played an important political role in forcing social partners to the negotiating table. Amazon warehouse workers there have organised more effectively due to this strong political support. At a 2023 summit about the impact of, and response to, Amazon, the former Italian Minister of Labour explained that, as a result of the government's mandate and union action, they were able to force Amazon to open social dialogue, leading to a strong union agreement including Amazon and their subcontractors.[23] The agreement secured was so strong that Italian workers have been leaving other jobs to join Amazon.

Attempts at downstream regulation attempt an entirely different proposition. As global and footloose as a production network might be, downstream regulation tackles the problem at a relatively static geography. While production can move from the Philippines to India or from South Africa to Kenya, the world's largest markets aren't going anywhere. An automobile firm will always want to sell their vehicles in France or Germany. A social media company will always want user engagement in the US. This simple fact gives policy-makers in those countries a potentially outsized role in setting standards.

When thinking about globe-spanning production networks, it might initially seem unrealistic that a law enacted in one country might somehow affect conditions of workers around the world. However, there are already a range of regulations on the books in a few countries which attempt to do exactly that.

In 2023, the German Supply Chain Law (with the very German-sounding name: Lieferkettensorgfaltspflichtengesetz) came into force. The law stipulates that all companies based in Germany with over 3,000 employees (from 2024, this number dropped to 1,000 employees) must ensure a range of minimum standards throughout their supply chain. In other words, by simply engaging with suppliers, those German companies have a legal responsibility to make them comply with the standards set out by the law, and to perform risk assessments to ensure compliance. Standards enshrined in the law include the need to identify, prevent or minimise the risks of child and forced labour, in addition to prohibitions on freedom of association, slavery, and a range of other human rights violations. While it is too early to evaluate the long-term effects of this law – and early signs indicate some considerable difficulties in applying it effectively to certain tech firms, it represents a significant reframing of the responsibility of lead firms throughout their production networks. Lead firms can no longer absolve themselves of moral responsibility for the conditions of workers on the other side of the world. The Supply Chain Law recognises that by coordinating a chain, a lead firm needs not just to ensure that products and services get produced to technical and operational specifications, but also to ethical ones.

A number of OECD countries have now implemented supply chain laws that work on improving business conduct via greater transparency and mandatory diligence. In Canada, Bill S-211 forces large Canadian companies to report on the steps that they have taken to reduce the risk of forced or child labour at any point in their production processes. The UK and Australian Modern Slavery Acts and the French Corporate Duty of Vigilance Law spell out similar requirements for firms based in those countries. The Norwegian Transparency Act goes a step further. In addition to mandating Norwegian companies to carry out due diligence activities throughout their supply chains, it also requires that they respond to information requests that enquire into working conditions in upstream parts of their supply chains.

With the exception of the German Supply Chain Law, most of these bills are focused primarily on modern slavery rather than a broader range of concerns relating to human rights and decent work. However, a proposed European Corporate Sustainable Due Diligence Directive has the potential to go much further (at the time of writing, the directive is advancing through the Trilogue process).[24] The Directive prohibits not only a range of environmental and human rights abuses, but also has provisions to classify remuneration below a local living wage, the prohibition of workers from joining a trade union, and unaddressed inequities in the workforce as violations. Like the earlier modern slavery-focused bills and the German Supply Chain Law, the extraterritorial scope of the bill means that the covered companies will not only have to be attentive to their own impacts, but will additionally need to monitor the impacts of all entities that are part of their supply chains and have any indirect business with it. The current draft of the Directive covers large EU-based companies and is also applicable to any firm (based anywhere on Earth) that generates a significant amount of revenue for the production and supply of goods or the provision of services in the EU. Therefore, this broad geographic remit could go a long way to preventing the offshoring of unethical practices and pay below living wages to parts of the world where regulators feel pressured by the competitive dynamics of the planetary market.

The Directive is highly unlikely to become a silver bullet and there are a number of important critiques that can be levelled at it. Its floors might become ceilings and might disincentivise companies from climbing above narrow compliance measures. It might prevent member states from legislating more ambitious standards. And it will undoubtedly contain numerous exceptions – for instance, excluding the finance and service sectors. Most importantly, provisions like the living wage seem unlikely to be applied to the self-employed or day labourers.

Irrespective of the gaps that the final text of the Directive will have, what really matters here is that it, and the assorted modern slavery and supply chain laws that preceded it, show that downstream regulation can be constructed to be extraterritorial: thus ensuring that addressing harms that might accrue to workers and the environment in any production process are not left to the ethical compass of corporate managers, but are rather imposed as statutory requirements. Were such a law to be passed in, say, Albania, Laos or Jamaica, it would be easy for large firms to ignore. However, the countries proposing these bills account for a significant portion of the world's exports in goods and services. Large tech firms will be unlikely to want to fall afoul of them and thus be excluded from those markets.

There will undoubtedly be many future battles to be fought about exactly how these laws and directives are configured and who they cover, but for corporations who coordinate global production networks they are undoubtedly the thin end of the wedge. We should be under no illusion that most states have precious little interest in protecting the lives, health and wellbeing of workers in other countries. However, these regulations may become sufficiently institutionalised and normalised that it could be hard to imagine going back to a world in which responsibility stops at national boundaries.

Finally, it is worth considering the role that network-wide regulation can play in taming the sector. Here the primary actor worth looking to is the International Labour Organisation (ILO). The ILO is an agency of the United Nations tasked with setting labour standards and promoting decent work. Its purview therefore stretches to 187 member states, which encompass the vast majority of humanity. There are no simple pathways to developing ILO conventions. However, once they are ratified by member states they become legally binding instruments. Should upstream and downstream regulation prove insufficient to prevent harms from being perpetuated in AI's production networks, an ILO convention could be useful to set minimum global standards for all workers in this sector.[25]

Step Four: Worker Cooperatives

Legislation protects workers by creating minimum standards and binding rules that employers must adhere to in how they treat their workers. In addition, trade unions can negotiate on behalf of workers and try to get them a fair share of the pie, but what if instead of fighting for a piece of that pie, workers bake their own? This is the task of worker cooperatives, organisations in which workers jointly own and manage the business. Imagine a bakery where the sixteen staff members who work there all have an equal financial stake in the business and get together regularly to jointly decide how it should be run. At the end of the day, anything they earn is distributed between them. Worker ownership is not only a way of building collective power and sharing the spoils of business, but also of exploring meaningful ways of implementing workplace democracy. This model might be less familiar to some readers in the US and UK, as cooperatives are more prominent in other countries. Cooperatives Europe, for example, reports that over 17 per cent of Europe's population are members of a cooperative.[26] Various forms of agricultural and banking cooperatives are also common across different parts of India, Asia and Africa.[27] The history of modern cooperativism can be traced back to the beginning of the Industrial Revolution and early visionary thinkers such as Robert Owen and Charles Fourier, with the first widely recognised cooperative founded in Rochdale, England in 1844 by twenty-eight cotton weavers.[28]

What are the prospects for workers' cooperatives in the AI production network? One example of a data annotation firm aspiring to a similar model is Karya, a non-profit based in the southern Indian city of Bengaluru, launched in 2021 and marketing itself as 'the world's first ethical data company'. The majority of the non-profit's work involves enabling people in central and southern India to create local-language speech datasets that are not currently well-covered in digital databases. In practice, it assists individuals recording themselves speaking their language into an app and creates new datasets that can be sold to tech companies.

Karya's origins stretch back to its three co-founders, two of whom worked at Microsoft Research. They were part of teams working on localising AI solutions to Indian languages. Of the 1.3 billion inhabitants of India, only 11 per cent speak English; and there are thirty other languages spoken by over a million native speakers. Many of those languages do not have much published in them, or about them, and there was thus an enormous data gap in South Asian languages. You can tell a virtual voice assistant to play a song or turn the lights off in English. But what if you wanted to do that in Kashmiri, Gujarati or Assamese? For many localised AI solutions to effectively work, voice data needs to be collected from native speakers to train models. Underpinning every voice-activated service are countless native language workers who have submitted training data.

Safiya Husain, one of the co-founders of Karya, observes that, 'AI training data is among the most lucrative assets on the planet, but very little of this money goes back to the people who actually do the work.' The founders calculated that median hourly wages of those workers were effectively about 10–50 cents, whereas the datasets they were producing were being sold for over 200 times that price.[29] Workers were keeping only half a per cent of the value they were creating!

Through its work, Karya enables workers to earn roughly $5 an hour, which is many times the national minimum wage in India. Karya limits the total number of hours a worker can work on the app per week, as the founders did not want it to become a full-time grind, but for the hours they do work, the wages are decent. High wages alone would not entail that we could consider Karya a full workers' cooperative, but there are elements of their model that push in this direction. The first is that the company uses a Public Data License that enables the workers who participated in creating a dataset to exercise collective ownership over it and continue to benefit from its sale to multiple companies. This concept draws on the idea of 'data cooperatives' – organisations that were formed to negotiate between data producers and companies seeking to use their data.[30] There are different ways in which this could be arranged, but creating a data cooperative could involve this organisation acting as a 'fiduciary intermediary' to rebalance power in the digital economy and give previously isolated data producers an opportunity to receive their share of dividends from different companies' use of their data. In Karya's case, not every dataset ends up getting resold, but some have been resold two or three times; sometimes even for higher amounts. In one example, as a result of datasets being resold, workers earned about 1,000 rupees an hour for their work (equivalent to about $12, and about forty times the national minimum wage).

In this instance, the 'data cooperative' model has been tailored to the Indian context. The 32,000 workers on the platform perform all of their tasks on phones. Karya works with local NGOs to recruit workers: ensuring gender and religious diversity in their worker pool, and ensuring that poor communities and marginalised castes can find jobs on their platform. Workers do those jobs from their homes, and from rural areas – without having to travel into cities. Because of this arrangement, there are limits to the extent this data cooperative could ever institute more formal democratic procedures for their workers to have a say in how the organisation is managed. It is difficult to imagine how thousands of workers could have the same influence or power over the organisation as its three founders. When we spoke to Safiya, there were plans to experiment with a workers' council, consisting of worker representatives who would have a formal role in decision-making. However, this idea was still in an early stage of development.

How scalable are cooperatives across the AI production network? One oft-noted limitation with cooperatives is the difficulty in raising sufficient capital for large-scale projects. Traditional venture capital and large institutional investors tend to shy away from cooperatives because shares do not give them a say in how the business is managed and returns on investment are generally smaller and take more time than traditional businesses. This lack of access to capital makes it tricky for these businesses to scale and compete against corporate rivals. Trebor Scholz, director of the Platform Cooperativism Consortium, argues that cooperatives work best when embedded in a supportive ecosystem of other cooperative businesses, investment institutions such as cooperative banks, and a regulatory framework with favourable tax incentives.[31] But these requirements can make it hard for lone cooperatives to develop and make it on their own. When it comes to capital-intensive investments in large data centres or AI labs that train foundation models, it is hard to see cooperatives having a realistic economic chance of survival. But for data annotation services, art collectives and various forms of data cooperatives that provide services to consumers, cooperatives are a concrete possibility that provide benefits to their workers.

The underlying point of cooperative businesses is that they contribute to a more plural and democratic economy. Their benefits should not only be seen in the financial returns they deliver to workers, but in the broader range of social goods, such as providing better services for stakeholders, deepening relationships between people and allowing people to exercise democratic governance in organisations that are meaningful to them. Although cooperatives are currently only a small part of the global economy, through federations of autonomous but connected cooperatives there is an opportunity for them to continue to grow. But when cooperatives come up against cutthroat corporations willing to exploit their workers, we see there are limits to how big they can grow, which highlights some of the problems that any ethical alternatives face within a broader capitalist economic system. What do you do when it's the whole damn system that is to blame?

Step Five: Dismantle the Extraction Machine. Build the Future

It was not that long ago that slavery and serfdom were widespread. Under those systems of economic organisation, workers were seen as a resource from which labour could be freely extracted. The idea that workers might have meaningful rights, such as autonomy over their time (and bodies), health and safety protections, and a living wage was inconceivable to many. Today, while remnants of those systems continue to exist, there are a suite of rights and protections that exist for workers all over the world. Those rights and protections are partial, full of gaps, and certainly insufficient to guarantee a dignified life for everyone, but they do nonetheless illustrate how much has changed. Those changes have happened neither by accident nor because of the benevolence of those who controlled history's extraction machines. Rights and protections have been won because people throughout history have demanded change.

Workers have progressively transformed the world by demanding more. They have brought us to where we are today by fighting for rights their parents and grandparents did not have. Because of those fights, a majority of the world can enjoy universal suffrage, two-day weekends, and the right to organise collectively. But these rights are highly geographically constrained. An eighteen-year-old in Switzerland or Sweden has many more guaranteed rights than their equivalent in Bangladesh or Burundi. In all countries, this journey towards greater fairness and dignity for workers is still far from complete. There is much that needs to be done, and the journey is one that we must all contribute towards.

This book has shed light on the complex global production networks of AI. The structure of these networks connects workers and rigs the system to create winners and losers before the game has even started. A Ugandan data annotator has no way of extracting any meaningful amount of value or concessions out of the system. They have no way to break the cycle of extraction.

At the dawn of the age of connectivity, it was common to think that access to information would be democratised and that a global 'cyberspace' would equalise opportunities and allow people from every corner of the planet to share in the world's knowledge. Those hopes were widespread because, in theory, that is exactly what digital tools allow us to do. They allow us to freely share information. They allow us to communicate instantaneously from one corner of the planet to another. Why didn't the digital revolution act as a great equaliser?

The answer is global capitalism. In a world in which there are enormous inequalities in knowledge, resources and capabilities, connectivity can be used to bring ever more of the world into an extraction machine. The AI industry is just the next phase in a long journey that stretches back to the age of colonialism. Today's extraction machine attempts to neuter our ability to rewire how it works. It sets up a system in which capital alone has the power to see all of the nodes in a planetary network of production, and it alone can command the spatial division of labour within that network.

If we are to build a fairer future, the solution is therefore not more connectivity or more equal access to technology. The solution is to dismantle the machine and build something else in its place. Just as kings and emperors did not willingly surrender their power, the rulers of today's digital empires are unlikely to choose to empower the disempowered. They will only do so if their hands are forced.

The powerful and the disempowered are all part of the same system; all part of the same extraction machine. So, if we are committed to a more just system, we have to think about how to rewire the machine itself. It would be dishonest to pretend that there is an easy, straightforward and simple pathway to doing so. But it is nonetheless essential to recognise the task ahead, to be ambitious, bold and creative in our demands and our expectations. If we can do that, we will find ways of collectively working towards a fair and just future.

Conclusion

Following an exchange of missiles and air strikes with Hamas during its eleven-day war in Gaza in 2021, the Israel Defense Forces (IDF) declared it had conducted its 'first AI war'.[1] Machine learning tools were at the heart of a new centre established in 2019 called the Targets Administrative Division that used available data and artificial intelligence to accelerate target generation. Former IDF Chief of Staff Aviv Kochavi said, 'in the past we would produce fifty targets in Gaza per year. Now, this machine produces one hundred targets [in] a single day, with 50 per cent of them being attacked.'[2]

Technology has long been sold as a tool to make war more humane. As early as the 1980s, when the IDF invaded and occupied Southern Lebanon, it was using the term 'surgical precision' to describe attacks by its air force.[3] Drawing from this rhetorical playbook, the United States began referring to precision-guided munitions as 'smart bombs' during the first Gulf War.[4] The stated goal was to reduce civilian casualties, but in the background was a PR strategy to sell war to the public as something other than mass slaughter. The IDF's AI-based targeting system can make it appear like targets are now selected with machine-like precision to minimise the indiscriminate use of force. In reality, the precise opposite is true.

In November 2023, an investigation by +972 Magazine and Local Call revealed the existence of a program called 'Hasbora' (the Gospel) that produces AI-generated targets and focuses not only on military targets but also on private residences, public buildings and high-rise residential blocks – so-called 'power targets' that might 'create a shock' to Palestinian civil society and lead them to exert more civilian pressure on Hamas.[5] 'The perception is that it really hurts Hamas when high-rise buildings are taken down, because it creates a public reaction in the Gaza Strip and scares the population,' said a source from the investigation. 'They wanted to give the citizens of Gaza the feeling that Hamas is not in control of the situation. Sometimes they toppled buildings and sometimes postal service and government buildings.'[6]

Following the 7 October 2023 attack by Hamas on Israel, the Israeli army loosened constraints on civilian casualties and began an AI-assisted bombing campaign. The +972 Magazine/Local Call investigation revealed that the Israeli army has records of precisely how many civilians are likely to be killed in an attack on particular targets and makes calculated decisions about what could be justified. An IDF spokesperson stated their goal is the complete eradication of Hamas and that 'while balancing accuracy with the scope of damage, right now we're focused on what causes maximum damage'.[7] In previous operations, the homes of junior Hamas members had not been targeted for bombings, but AI now expanded the range of targets possible for attack. 'That is a lot of houses,' an Israeli official told +972/Local Call investigators. 'Hamas members who don't really mean anything live in homes across Gaza. So they mark the home and bomb the house and kill everyone there.'

According to an IDF statement on 11 October 2023, 1,329 out of a total 2,687 targets bombed were deemed 'power targets', which could include 'high-rises and residential towers in the heart of cities, and public buildings such as universities, banks, and government offices'.[8] Further +972 investigations revealed that during these early days of Israel's genocidal assault the IDF used 'kill lists' generated by an AI targeting system known as Lavender to attack up to 37,000 targets with minimal human verification. The system uses mass surveillance data and machine learning to rank the likelihood that any one individual in Gaza is active in the military wing of organisations like Hamas from 1–100. Another system called Where's Daddy? was used to confirm the targeted individuals had entered their family homes before they were attacked, usually at night, by 'dumb' bombs that were guaranteed to cause 'collateral damage'. According to sources within the Israeli military, the result was thousands of Palestinian women and children dying on the say-so of an AI system.[9] In these cases, artificial intelligence was used to massively expand the capacities of a state military apparatus to conduct a war in which enormous civilian casualties resulted from the pursuit of putatively military ends. AI hasn't saved civilian lives; it has increased the bloodshed. Nor is AI limited to targeting programs; it is used across the Israeli military, including in another AI program called Fire Factory, which assists with the organisation of wartime logistics. As Antony Loewenstein has shown, once tested in combat on Palestinians, IDF military technology is then exported to conflict zones across the world by Israeli security companies.[10]

This example of the military application of AI relies on global production networks involving a hidden army of workers across the world. In this book, we have shown that these vast networks distribute decision-making power unevenly and are directed by powerful companies for their own benefit. For the most part, workers do not know what happens elsewhere in the network, with the whole system remaining opaque to all but a few coordinating actors. The Israeli military accesses its AI and machine learning capabilities via Google and Amazon, which provide cloud computing services for its operations in a controversial contract called 'Project Nimbus'.[11] After winning the Project Nimbus contract both Amazon and Google began spending hundreds of millions of dollars on new state-of-the-art data centres in Israel, some of them underground and secured against missile strikes and hostile actors. It's a long way from data-centre worker Einar and his home in Blönduós, Iceland, but these centres will employ other technicians like him to keep the system functioning. The award of these contracts also gave rise to new forms of worker organising, such as that undertaken by the Kenyan labour organiser, Paul. Hundreds of Google employees, part of the Jewish Diaspora in Tech group, protested the contract, signing a statement stating, 'Many of Israel's actions violate the UN human rights principles, which Google is committed to upholding. We request the review of all Alphabet business contracts and corporate donations and the termination of contracts with institutions that support Israeli violations of Palestinian rights, such as the Israel Defense Forces'.[12]

The military systems built for the IDF also rely on machine learning engineers with specialised knowledge and expertise to design the models and integrate them into Israel's existing military capabilities.

Israel's compulsory military service enables the army to recruit many of its brightest minds into its intelligence division and to train them as engineers and cybersecurity experts. Just as the US military-industrial complex played a key role in the early development of Silicon Valley, Israel's military and defence industry is a launch pad for software engineers and high-tech startups.[13] Many Israeli entrepreneurs serve in the military, where they have direct experience with Israel's sophisticated system of computerised surveillance and other technologies of war used as part of the occupation. When it comes to computer vision systems such as military targeting technology, facial recognition software and autonomous drones, images and videos need to be curated and annotated by an army of data annotators, many of whom are employed via outsourcing centres such as Anita's in Gulu, Uganda. This hidden human workforce powers these AI systems and remains trapped within a global network that maintains unequal access to capital, resources and work opportunities.

We have also highlighted how this AI production network is fundamentally colonial in character – many of its connections can be traced back to colonial origins and continue to be shaped by the historical legacy of colonialism today. At a very basic level, we have seen how networks of connection literally trace the same paths as older shipping routes and telegraph cables of former colonial empires. But more importantly, these production networks ensure that value and resources flow from peripheral nodes of the network to the centres of accumulation. Countries with advanced high-tech economies at the centre of global capitalism use their power to maintain their wealth and status within the system by leading new ventures that extract value from the labour, critical minerals and data of populations in peripheral countries.[14] This can be seen at a very visceral level in the global distribution of work that produces AI: secure and well-paid jobs of highly trained engineers and technicians are maintained at the centre of these networks, while low-paid and menial tasks are outsourced to countries with permissive labour regulations and a more vulnerable workforce in countries in the Global South. These networks turn these countries into sites of extraction and exploitation and in this sense continue older patterns of colonial power that led to these countries being dispossessed and underdeveloped in the first place.

But ultimately, we argue these production networks are porous, potentially malleable and able to be challenged and transformed. The way we produce AI is not set in stone. Nor is it the result of some malevolent force over which we have no control. Real people are making choices. They respond to structural pressures from the market and a desire to enhance their position within the network, but this does not mean that nothing can be done to change the system. These networks are changing all the time and responding to pressures at different points in the supply chain. What happens to workers at one point can have profound consequences for those at another. There are critical junctures where organised workers have the power to make a difference. The first step is to understand how these systems work so we know how to change them.

Technological development can often have a deterministic feel to it. We are told that we should adapt to the inevitable changes to our world brought about by new technologies rather than have a say in how they are developed. The reality is that the world of tech is a highly political and contested field. We can collectively shape a different kind of future for AI. Prophecies of what will come to pass made by entrepreneurs with investments in tech stocks are much more about willing a new world into existence for the sake of their balance sheets than making logical and dispassionate predictions of what might happen. But these billionaire-funded projects do not always succeed. Despite what tech companies would have us believe, the future is still very much up for grabs.

Before we outline what we can all do to shape this future, we have some key messages that we'd like to be at the forefront of your mind when you close this book.

1. *AI data annotation work is part of a planetary labour market in which work can be shifted at a moment's notice anywhere in the world. Workers must organise transnationally to defend their interests.*

We have seen how workers in AI production networks compete in a global market characterised by a distinct spatial division of labour. The lower-skill data work in those networks is relatively footloose and can – in theory – be done from just about anywhere on Earth. Employers take advantage of this global system to pit workers from different parts of the world against one another, to push for ever-cheaper labour costs and to remind workers that lobbying for better wages might even be against their own interests. In this market, it is extremely hard for any isolated group of workers in a particular geographic location to unilaterally lobby for better conditions. Companies in the Global North can set the terms of the contract and force outsourcing centres to compete against one another for their business. This race to the bottom only serves the companies that facilitate it.

The complexity and globe-spanning nature of AI production networks mean that workers will have to find ways of organising across national boundaries if they wish to challenge the power of large corporations. Workers have to find new ways of organising across supply chains and standing in solidarity with workers at other points in the network. Tech companies have a natural advantage in already being well-resourced actors with significant agenda-setting power and leverage over how these networks operate. Workers must organise collectively and act together to build countervailing sources of power and put pressure on companies to improve pay and conditions. It is only worker-led organisations that will properly defend their interests.

2. AI-driven work management systems are coming for you.

The real power of AI when it comes to work is its ability to intensify and deskill work processes through increased surveillance, routinisation and more fine-grained control over the workforce. AI allows bosses to track workers' movement, monitor their performance and productivity, centralise more data about the labour process in managerial hands, and even claims to detect their physical and emotional states. It is also increasingly used by HR departments for 'hire and fire' decisions – screening applicants before CVs are seen by a human and triggering disciplinary and termination procedures when targets are not met. Gig work has been the canary in the coal mine for this type of algorithmic management, but the technology is quickly spreading to a range of other forms of work.[15] If you think your job is immune, you are probably mistaken. Once tested on workers with weaker bargaining power, these technologies are then rolled out to broader sectors. There they are used to cut costs either by replacing functions previously performed by humans, or more often, by increasing the pace at which humans must work and reducing the skills required to perform a specific job.

AI management technology is overwhelmingly designed for the benefit of managers and owners, not workers. As a result, work intensity can increase to unsafe levels, and deskilling can reduce workers' autonomy and job quality. The result is a widespread tendency across all sectors of the economy towards tiring, dehumanising and dangerous work. The Amazon worker who stands in one place repeating the same small tasks thousands of times a day to make the rate that their AI manager dictates is experiencing this future of work. They are one of the millions of workers who are exposed to a high risk of injury, subject to tracking, and pushed to work harder and harder every day. From the data produced by their scanning guns to the patterns observed by cameras overhead, they experience AI management as a constant pressure to perform.

Beyond the warehouse, the pandemic provided an opportunity for employers to start using these monitoring techniques on employees' productivity as they worked from home. Microsoft was criticised for offering 'productivity scores' in one of their software suites that could have been used to allow managers to track how actively employees contributed to activities such as emailing and collaborative documents.[16] The use of cameras to monitor workers in the office, keystroke and computer activity monitoring and software that tracks and records performance is becoming more widespread. Nobody is exempt from this future of work.

3. *AI will concentrate power into fewer hands and change the dynamic between workers and their bosses.*

The AI arms race risks consolidating the power over this technology into just a handful of firms and individuals that could wield control over decisions about how AI is developed for billions of people across the world. As we have seen, training foundation models require large amounts of computing power and vast repositories of data. These resources are tightly controlled by the largest tech companies, which exercise infrastructural power through ownership and control over data centres and compute resources. The capital-intensive nature of infrastructure like hyperscale data centres hands the big players in the market yet another advantage. A report from the research non-profit, AI Now Institute, points out that 'the majority of existing large-scale AI models have been almost exclusively developed by Big Tech, especially Google (Google Brain, Deepmind), Meta, and Microsoft (and its investee OpenAI)'; just as AI cloud computing is also 'a market already concentrated in Big Tech players, such as AWS (Amazon), Google Cloud (Alphabet), and Azure (Microsoft)'.[17] These companies are also some of the most profitable in the world, with enormous valuations and large cash reserves with which to invest in and gain influence over younger startups and AI labs. This allows them to integrate new technology into their existing products and dominate the AI market.

The potential downsides are particularly exaggerated for workers. In the wider economy, AI products controlled by a select few companies are developed to maximise efficiency and squeeze every bit of work out of them, forcing them to work faster and harder through increased surveillance and performance monitoring software. The asymmetrical deployment of AI tools is likely to further reinforce the division of the economy into two camps: the big monopolies and the zombie firms. The growing power of big firms in the workplace and marketplace suggests a depressing future ahead for a workers' movement that has already been decimated by neoliberal reforms and a hollowing out of institutions for collective bargaining and the representation of workers' interests. AI risks further cementing the power of capital against workers and placing workers in competition with each other.

But no level of technological development in the history of capitalism has yet eliminated workers' capacity to self-organise. As AI management has started to proliferate, strike numbers in the UK and US have spiked upwards from historic lows. AI has been a feature of many of these disputes, with the Communication Workers Union branding the technology as one of the most significant threats facing their workers in the UK postal system.[18] Inside the AI production networks that create this management technology, the concentration of power in the hands of a few firms creates new opportunities for workers. Historically, waves of consolidation and centralisation within industries have created better conditions for unionisation as workers face a common enemy, rather than being divided across different firms.[19] The same trend might repeat itself now. The new dynamics of AI-managed work will ultimately be determined by the relations between workers and managers that prevail on the floor of warehouses and offices across the world.

4. AI is not akin to a nuclear weapon or a future extinction event – the risks it poses are real and present today.

AI is a cutting-edge technology that has developed quickly and poses large and unpredictable risks to society. This is where the comparison with nuclear weapons ends. It's also how many new dangerous technologies could be described, making it hard to distinguish between them. The problem with such comparisons is they conjure unhelpful analogies of *Terminator*-style 'AI takeover' events, which by their very nature are based almost entirely on abstract speculation of where we might be twenty or a hundred years in the future. But when we think about actually existing AI – computer vision systems, facial recognition technology, autonomous vehicles, recommendation systems, virtual assistants and chatbots – there are numerous harms that are already happening right now. Forget about future synthetic autonomous agents – humans in the present are more than capable of using AI for nefarious purposes – to run misinformation campaigns, create deepfakes, interfere in elections and harm others. On top of this, the very design of these systems often reproduces biases present in human societies in a new algorithmic form. As AI companies race forward with developing frontier technologies, we should of course be mindful of how powerful models are becoming, but a violent coup on behalf of autonomous AI systems would require technology so fundamentally different from our current generation of AI that alarmist concerns do not contribute productively to a realistic assessment of the dangers it poses today. AI is being used to improve the lethality of current militaries, including in Gaza, without any autonomous agents in sight. How we respond to these harms in the present will help shape the medium- and far-term future in ways that will help us prepare for future technological developments.

5. *If AI is understood as an extraction machine, then we are the raw material.*

If this book could be distilled down to a single message it is that we, human beings, are the often-hidden force that powers AI – both physically with our labour but also intellectually through AI ingesting and synthesising our collective intelligence. We are the human hidden in the wooden box that enables the Mechanical Turk to perform genius chess moves, seemingly automatically. Without us, AI ceases to function. It is only through the constant supply of human labour – annotating datasets, coding software, repairing servers, creating new paintings and literature and keeping supply chains functioning – that AI continues to exist. The extraction machine uses human beings like raw material, churning through ever-larger datasets and quantities of knowledge to power its algorithms. As we have shown, the machine makes use of workers in different ways depending on their positions in global capitalism. But they are united in all being directed by the logic of extraction. The machine has a purpose: to enrich tech company shareholders and concentrate power in the hands of a narrow elite.

What unites every character in this book – aside from the investor – is that in one way or another, their labour is just grist for the mill. It has been pointed out that data annotation work treats workers like machines, giving them repetitive tasks and closely monitoring their movements and performance. But the truth is even worse – these workers are treated as little more than the fuel needed to keep the machine running. They have a shared fate of being exploited by the extraction machine and used up in its quest to continue building itself up and growing at an exponential pace. Whether or not they can unite in opposition to it and the global regime that enforces these inequalities depends on their actions right now. In one way or another, many of us could be a character in this book. All of us who are fed into the extraction machine share a common fate.

How this narrative ends depends on what we do. It depends on indignant citizens, workers and consumers understanding how they are being used and that they can fight back. This is not an external struggle happening in a faraway land or a distant threat that will never appear before us. This is happening to you, and it's happening right now. In this book, we have told the stories of seven workers around the world who are caught up in the extraction machine. Mercy, the content moderator from Kenya who saw her grandfather killed in a video she was forced to watch over and over again; Anita, the data annotator from rural Uganda who builds the datasets that the world's largest companies rely on; Li, the London-based machine learning engineer who struggles with the ethical implications of the tech she's building; Einar, the Icelandic data centre technician who sits in the middle of a global web of infrastructural power; Laura, the actor who competes for jobs with an AI version of her own voice; Alex, the Amazon worker from a deindustrialised city who went on wildcat strike to try and win a pay rise; and Paul, who overcame overwhelming odds to found the first ever union for content moderators in Africa. These might sound like the stories of other people, who live very different lives to you in very different places – but they aren't.

Their story is also your story. Beyond just a basic level of human experience, you share a specific social position. On a probability basis alone, you are a member of the vast global majority of workers, consumers and citizens that the extraction machine treats as an exploitable resource. The machine wants your labour, your ideas, your art, your water, your energy, your data, and your country's critical minerals. All of these inputs are to be fed into the fire that converts these into outputs, power and profits.

There is a simple name for the system that has built this machine. It's capitalism, a social order built on the private ownership and control of the economy that systematically converts everything it touches into money. However much the venture capitalists of Silicon Valley and elsewhere have relied on governments for their research budgets, the process of technological development that led to this point has been driven by a capitalist system and led to capitalist outcomes. An isolated minority of investors and managers have taken the decisions, and the rest of us are living with the consequences. These consequences are not inevitable. It's incorrect to assume that any path of technological development would have produced identical challenges. The technology capitalism develops isn't neutral; it is built in the image of the system that birthed it.

The extraction machine shows its family lineage in some very obvious ways. It centralises control in the hands of a class of owners by rearticulating patterns of power formed during the colonial expansion of the capitalist economy. It allows its parent system to intensify and deskill labour, thereby squeezing more profit out of the work of an already overburdened global working class. It relocates production around the world to get access to the cheapest labour and the most readily available resources. Amid the anarchic swirl of hype and bubbles that characterises this kind of system, we can detect one rhythm emerging over and over again: a constant cycle of investment and profit. AI is not just a tool of global capital, it is integrated within it, as one of its newly developed but already essential organs.

We can imagine other uses for this technology: a world in which AI automates low-quality work, coordinates the use of scarce resources, or contributes to advances in scientific research. But this vision does not just require good intentions from key actors in the global production networks of AI. What is needed is something altogether more profound: a movement of that global majority capable of changing the social relations that have built the extraction machine. Our specific contributions to this movement will depend on where we are and what we do. We hope that in writing this book, we have helped you understand why we feel we cannot go on with business as usual. Your exact contribution is yet to be defined, but it will be based on the person you are, the place you find yourself, and the power you are capable of building alongside other members of that great global majority. Workers across the world are leading the way, starting to undertake the first struggles in what must ultimately be a long campaign. Engineers are refusing to build surveillance tech for governments, Amazon workers are walking out for a pay rise, and content moderators are banding together to demand better. Together, we can follow their lead to smash the extraction machine. Then, our task will be to combine the fragments into something more emancipatory – into a form of technology that accelerates the growth of human freedom.

On 2 December 1964, a student activist called Mario Savio gave a speech at the University of California, Berkeley. He was addressing a protest demanding that university administrators remove their restrictions on political speech and action on campus in the wake of police repression of civil rights activists. He began by telling the crowd how the university president had referred to the institution as if it were a corporation, and how, in this vision, the students were the raw material to be processed by the university. His voice swelled:

> There is a time when the operation of the machine becomes so odious, makes you so sick at heart, that you can't take part; you can't even passively take part, and you've got to put your bodies upon the gears and upon the wheels, upon the levers, upon all the apparatus, and you've got to make it stop. And you've got to indicate to the people who run it, to the people who own it, that unless you're free, the machine will be prevented from working at all.[20]

Savio's words apply just as well to our situation today as they did to the University of California sixty years ago. We too refuse to be the raw material that is fed into the extraction machine. We too are willing to put our bodies upon the gears of a system that chews up human labour and spits out profit. We too wish to indicate to the people who run it, to the people who own it, that unless we are free, the extraction machine will be prevented from working at all.

Acknowledgements

Some of the ideas in this book began to emerge during the first of many fieldwork trips to study Nairobi's digital outsourcing sector in 2009. Shortly after the first of many fibre-optic cables began to connect Kenya to the world's fibre-optic grid, the permanent secretary in Kenya's Ministry of Information and Communication, Professor Bitange Ndemo, generously welcomed Mark into his office in downtown Nairobi to discuss his vision for Kenya's place in the global economy. That discussion sparked a universe of further connections and conversations with digital economy stakeholders in Africa. We therefore wish to particularly thank Bitange for opening so many doors and offering introductions to the full spectrum of actors in the sector. It would not have been possible to conduct a decade and a half of research on digital work in Africa without Professor Ndemo's initial trust and enthusiasm.

That initial research, which began in 2009 based at the Oxford Internet Institute, was seed funded by the British Academy. It led to follow-on support from the ESRC and DFID to study the impacts of changing connectivity in East Africa, and then eventually a five-year European Research Council-funded project that Mark led called Geonet. The project brought together a skilled team of researchers focusing on Africa's digital economy. Over the course of those projects, Mark is indebted to a long list of collaborators who he worked with to study the outsourcing of digital work to the Global South. He would especially like to acknowledge Felix Akorli, Mohammed Amir Anwar, Fabian Braesemann, Chris Foster, Isis Hjorth, Charles Katua, Grace Magambo, Laura Mann, Sanna Ojanperä, Stefano De Sabbata, Fabian Stephany, Ralph Straumann, Tim Waema and Michel Wahome, for joining the research teams that he assembled and for their collaboration and support in planning research, collecting and analysing data, and for thinking through so many of the ideas in this book with him.

We have also deeply benefited from the conversations, debates and political engagements that the authors have carried out as part of the Fairwork project based at the Oxford Internet Institute and Berlin Social Science Centre. We would like to single out Funda Ustek Spilda for particular thanks. Funda helped design, execute, and think through much of our research project to study fair work in the production networks of AI. This would be a much-diminished project, and book, without Funda's crucial support and inputs.

Roberto Mozzachiodi, David Brand and Matt Cole also participated in the Fairwork AI project and also helped us collect data from Amazon workers and global AI stakeholders. Lola Brittain conducted an enormous amount of background research for us, and we are immensely grateful for her talents as a research assistant. Cheryll Soriano provided crucial contextual information about the BPO industry in Manila. Jonas Valente and Rafael Grohmann provided invaluable feedback on our first Fairwork AI report, which very much shaped our thinking about parts of this book. Finally, a big thank-you to all of the other Fairwork team members, who play such an important part in showing how research can really help to build a fairer future of work. Jana Ababneh, Eisha Afifi, Wirawan Agahari, Pablo Aguera Reneses, Iftikhar Ahmad, Tariq Ahmed, Shamarukh Alam, María Belen Albornoz, Luis Pablo Alonzo, Oğuz Alyanak, Hayford Amegbe, Branka Andjelkovic, Marcos Aragão, María Arnal, Arturo Arriagada, Daniel Arubayi, Sami Atallah, Tat Chor Au-Yeung, Ahmad Awad, Razan Ayesha, Adam Badger, Meghashree Balaraj, Joshua Baru, Ladin Bayurgil, Ariane Berthoin Antal, Alessio Bertolini, Wasel Bin Shadat, Virgel Binghay, Ameline Bordas, Maren Borkert, Álvaro Briales, Lola Brittain, Joe Buckley, Rodrigo Carelli, Eiser Carnero Apaza, Eduardo Carrillo, Maria Catherine, Chris King Chi Chan, Henry Chavez, Ana Chkareuli, Andrea Ciarini, Antonio Corasaniti, Pamela Custodio, Adriansyah Dhani Darmawan, Olayinka David-West, Luisa De Vita, Alejandra S. Y. Dinegro Martínez, Brikena Kapisyzi Dionizi, Ha Do, Matías Dodel, Marta D'Onofrio, Elvisa Drishti, Veena Dubal, James Dunn-Willimason, Khatia Dzamukashvili,

Pablo Egaña, Dana Elbashbishy, Batoul ElMehdar, Elisa Errico, Úrsula Espinoza Rodríguez, Patrick Feuerstein, Roseli Figaro, Milena Franke, Sandra Fredman, Farah Galal, Jackeline Gameleira, Pía Garavaglia, Chana Garcia, Michelle Gardner, Navneet Gidda, Shikoh Gitau, Slobodan Golusin, Eloísa González, Rafael Grohmann, Martin Gruber-Risak, Francisca Gutiérrez Crocco, Seemab Haider, Khadiga Hassan, Richard Heeks, Sopo Japaridze, Mabel Rocío Hernández Díaz, Luis Jorge Hernández Flores, Victor Manuel Hernandez Lopez, Nur Huda, Huynh Thi Ngoc Tuyet, Francisco Ibáñez, Neema Iyer, Tanja Jakobi, Athar Jameel, Abdul Bashiru Jibril, Ermira Hoxha Kalaj, Raktima Kalita, Revaz Karanadze, Zeynep Karlidağ, Lucas Katera, Bresena Dema Kopliku, Maja Kovac, Zuzanna Kowalik, Anjali Krishan, Martin Krzywdzinski, Amela Kurta, Ilma Kurtović, Morad Kutkut, Tobias Kuttler, Arturo Lahera-Sánchez, Jorge Leyton, Annika Lin, Georgina Lubke, Bilahari M, Raiyaan Mahbub, Wassim Maktabi, Oscar Javier Maldonado, Laura Clemencia Mantilla-León, Claudia Marà, Ana Flavia Marques, Margreta Medina, Rusudan Moseshvili, Jamal Msami, Karol Muszyński, Hilda Mwakatumbula, Beka Natsvlishvili, Mounika Neerukonda, Ana Negro, Chau Nguyen Thi Minh, Sidra Nizambuddin, Claudia Nociolini Rebechi, Bonnita Nyamwire, Mitchelle Ogolla, Oluwatobi A. Ogunmokun, Frederick Pobee, Daviti Omsarashvili, Caroline A Omware, Nermin Oruc, Christian Nedu Osakwe, Balaji Parthasarathy, Francesca Pasqualone, María Inés Martínez Penadés, Leonhard Plank, Frederick Pobee, Valeria Pulignano, Jack Linchuan Qiu, Jayvy R. Gamboa, Ananya Raihan, Antonio Ramírez, Juan-Carlos Revilla, Ambreen Riaz, Alberto Riesco-Sanz, Nagla Rizk, Moisés K. Rojas Ramos, Federico Rosenbaum Carli, Cheryll Ruth Soriano, Julice Salvagni, Derly Yohanna Sánchez Vargas, Maricarmen Sequera, Murali Shanmugavelan, Aditya Singh, Shanza Sohail, Janaki Srinivasan, Anna Sting, Isabella Stratta, Zuly Bibiana Suárez Morales, David Sutcliffe, Mubassira Tabassum Hossain, Tasmeena Tahir, Ainan Tajrian, Dinh Thi Chien, Kiko Tovar, Funda Ustek Spilda, Jonas Valente, Giulia Varaschin, Daniel Vizuete, Annmercy Wairimu, Jing Wang, Robbie Warin, Nadine Weheba, Katie J. Wells, Anna Yuan and Sami Zoughaib. The Fairwork project would not be possible without the support of the German Federal Ministry for Economic Development and their cooperation partners at GIZ. At GIZ, we wish to particularly thank Shakhlo Kakharova, Kirsten Schuettler and Lukas Sonnerberg for their boundless support for our work.

Both Mark and Callum benefited from being working group members of the GPAI Future of Work Working Group. Thank you to everyone at GPAI for creating such a stimulating environment to think through the implications of AI in the future of jobs.

Our work found a supportive home at the Oxford Internet Institute, and we wish to thank Duncan Passey, Michelle Gardner, Victoria Nash, Katia Padvalkava and Joanna Barlow for their extensive administrative support. We thank David Sutcliffe for providing editorial support on Chapter 7. Essex Business School also provided essential support for James and Callum throughout the writing process. In particular, they would like to thank Professor Peter Bloom for his tireless work in facilitating a unique research environment. Some of the work in this book was also conducted while Mark was a visiting researcher at the Berlin Social Science Centre. A big thank-you to Martin Krzywdzinski and his team for being such warm and generous hosts and for providing such an intellectually stimulating environment to think through the globalisation of work.

Thank you to everyone at the Autonomy think tank for providing a supportive environment for this research project to be undertaken and for fostering the development of new ways of thinking about technology.

Thank you to Bernie Hogan for providing such thoughtful feedback to our arguments about intelligence in Chapter 2.

James would like to thank his family, including Yasamin, Catriona, Peter, Michael, Stacey, Noah, Elizabeth and Sarah for all their love and support over the years. The book would also never have been written without cuddles from his two mini-dachshunds, Barcus and Karly.

Mark would like to thank: his mum, Jean Graham. As someone who works on digital labour platforms, she has helped him understand the precarity of a planetary labour market first-hand; his dad, Hashem Hashemi, for the endless words of encouragement. They really mean a lot! And to Caroline for being such a rock of support. Thank you.

Callum would also like to thank: his mum, Judy Cant, for her relentless belief; his dad, John Cant, for both his comments on Chapter 5 and insights into how technical systems really work; his wife, Evelyn Gower, for being his teammate and laughing at his worst jokes; and his co-editors of *Notes From Below*, for continually renewing the method of workers' inquiry that guides his research.

A big thank you to Will Francis and Corissa Hollenbeck at Janklow & Nesbit, Ben Hyman at Bloomsbury, and Simon Thorogood and Fraser Crichton at Canongate for their trust, encouragement and support as we turned our initial ideas into this manuscript.

This book draws on ideas we have developed over years of research and activism with groups of workers all over the world. It is impossible to understand the stakes of AI in our contemporary context without this engagement: trying to do so is like studying maths without using the concept of zero. From rainy street corners to picket lines and pubs, we have been privileged to engage with the collective intelligence of workers who see the reality of technological development and deployment first-hand every day. Their time, openness and insight have been an essential precondition for this book: without them, there would be no *Feeding the Machine*. So, our final thanks go to them.

Notes

Introduction

1   The decisions taken by human content moderators also serve to build training datasets that are used by firms like Meta to build AI content moderators. These systems perform basic content moderation tasks, like prioritising the most sensationalist content, detecting duplicates, and making basic decisions on content that is highly likely to violate the platform's terms of service.

2   Our definition draws from the OECD's updated definition of artificial intelligence. See Bertuzzi, Luca. 'OECD updates definition of Artificial Intelligence "to inform EU's AI Act"'. *Euractiv*, 9 November 2023, www.euractiv.com/section/artificial-intelligence/ news/oecd-updates-definition-of-artificial-intelligence-to-inform-eus-ai-act/.

3   Thormundsson, Bergur. 'Artificial intelligence (AI) market size worldwide in 2021 with a forecast until 2030'. Statista, 6 October 2023, www.statista.com/statistics/1365145/artificial-intelligence-market-size/.

4   Crawford, Kate. *Atlas of AI*. Princeton: Princeton University Press, 2021.

5   Gray, Mary L. and Suri, Siddharth. *Ghost Work: How to Stop Silicon Valley from Building a New Global Underclass*. Boston, MA: Houghton Mifflin Harcourt, 2019; Tubaro, Paola, Casilli, Antonio A. and Coville, Marion. 'The trainer, the verifier, the imitator: Three ways in which human platform workers support artificial intelligence'. *Big Data & Society*, Vol. 7 No. 1 (2020), doi.org/10.1177/2053951720919776; Williams, Adrienne, Miceli, Milagros and Timnit Gebru. 'The Exploited Labour Behind

Artificial Intelligence', *Noēma*, 13 October 2022, www.noemamag. com/the-exploited-labor-behind-artificial-intelligence/.

6   Irani, Lilly. 'The cultural work of microwork'. *New Media & Society*, Vol. 17 No. 5 (2015), pp. 720–39, doi.org/10.1177/ 1461444813511926.

7   Leswing, Kif. 'Meet the $10,000 Nvidia chip powering the race for A.I.'. *CNBC*, 23 Februrary 2023, https://www.cnbc. com/2023/02/23/nvidias-a100-is-the-10000-chip-powering-the-race- for-ai-.html.

8   Kak, Amba and West, Sarah Myers. 'AI Now 2023 Landscape: Confronting Tech Power'. *AI Now Institute,* 11 April 2023, ainowinstitute.org/2023-landscape.

9   Zuboff, Shoshana. *The Age of Surveillance Capitalism*. New York: Public Affairs, 2019.

10   Quijano, Anibal and Ennis, Michael. 'Coloniality of Power, Eurocentrism, and Latin America'. *Nepantla: Views from South*, Vol. 1 No. 3 (2000), pp. 533–80, www.muse.jhu.edu/article/23906; Maldonado-Torres, Nelson. 'On the Coloniality of Being'. *Cultural Studies*, Vol. 21 No. 2–3 (2007), pp. 240–70, doi.org/10.1080/ 09502380601162548.

11   See Muldoon, James, and Wu, Boxi A. 'Artificial Intelligence in the Colonial Matrix of Power'. *Philosophy & Technology*, Vol. 36 art. 80 (2023), doi.org/10.1007/s13347-023-00687-8. This argument has also been made by a number of other writers, for example, Mohamed, Shakir, Png, Marie-Theres, and Isaac, William. 'Decolonial AI: Decolonial Theory as Sociotechnical Foresight in Artificial Intelligence'. *Philosophy & Technology*, Vol. 33 (2020), pp. 659–84, doi.org/10.1007/s13347-020-00405-8.

1: The Annotator

1   Branch, Adam. 'Gulu in War and Peace? The Town as Camp in Northern Uganda'. *Urban Studies*, Vol. 50 No. 15 (2013), pp. 3152–67, doi.org/10.1177/004209801348777; Human Rights Focus, *Between two fires: the plight of IDPs in northern Uganda.*

Gulu, 2002, https://www.humiliationstudies.org/documents/
OnenBetweenTwoFires.pdf.

2   Perrigo, Billy. 'Exclusive: OpenAI Used Kenyan Workers on Less
Than $2 Per Hour to Make ChatGPT Less Toxic'. *Time Magazine*,
18 January 2023, time.com/6247678/openai-chatgpt-kenya-workers/;
Dzieza, Josh. 'AI Is a Lot of Work'. *New York Magazine*, 20 June
2023, nymag.com/intelligencer/article/ai-artificial-intelligence-
humans-technology-business-factory.html.

3   See Whittaker, Meredith. 'Origin Stories: Plantations, Computers,
and Industrial Control'. *Logic(s) Magazine*, n.d, logicmag.io/supa-
dupa-skies/origin-stories-plantations-computers-and-industrial-
control/.

4   Van der Linden, Marcel. 'Unfree Labour: The Training Ground
for Modern Labour Management'. In: Marcel van der Linden,
Global Labour History: Two Essays. Noida: V.V. Giri National
Labour Institute, 2017, pp. 13–27, vvgnli.gov.in/sites/default/
files/125-2017-%20Marcel%20van%20der%20Linden.pdf;
Menard, Russell R., *Sweet Negotiations: Sugar, Slavery, and
Plantation Agriculture in Early Barbados*. Charlottesville:
University of Virginia Press, 2006; Morgan, Philip D. 'Task and
Gang Systems: The Organization of Labor on New World
Plantations'. In Stephen Innes (ed.), *Work and Labour in Early
America*. Chapel Hill: University of North Carolina Press, 1988,
pp. 189–222.

5   See Rosenthal, Caitlin. *Accounting for Slavery: Masters and
Management*. Cambridge, MA: Harvard University Press, 2018;
Manjapra, Kris. 'Plantation dispossessions: The global travel of
agricultural racial capitalism'. In: Beckert S., Desan, C. (eds.),
American Capitalism: New Histories. New York: Columbia
University Press, 2018, pp. 361–88.

6   Taylor, Frederick W. *The Principles of Scientific Management*. New
York: Harper, 1911.

7   Ouma, Stefan, Premchander, Saumya. 'Labour, Efficiency, Critique:
Writing the Plantation Into the Technological Present-Future'.
Environment and Planning A: Economy and Space, Vol. 54 No. 2

(2021), pp. 413–21, ideas.repec.org/a/sae/envira/v54y2022i2p413-421. html.

8   Griggs, Troy and Wakabayashi, Daisuke. 'How a Self-Driving Uber Killed a Pedestrian in Arizona', *New York Times*, 21 March 2018, www.nytimes.com/interactive/2018/03/20/us/self-driving-uber-pedestrian-killed.html.

9   Cloudera. AI Data Engineering Lifecycle Checklist, www.cloudera.com/content/dam/www/marketing/resources/whitepapers/ai-data-lifecycle-checklist-cloudera-whitepaper.pdf.

10   Jones, Phil. 'The Mechanical Turk', *Verso Blog*, 26 November 2021, www.versobooks.com/en-gb/blogs/news/5223-the-mechanical-turk.

11   Gran View Research, Data Collection And Labeling Market Size, Share & Trends Analysis Report By Data Type (Audio, Image/Video, Text), By Vertical (IT, Automotive, Government, Healthcare, BFSI), By Region, And Segment Forecasts, 2023–2030, www.grandviewresearch.com/industry-analysis/data-collection-labeling-market.

12   Pontin, Jason. 'Artificial Intelligence, With Help From Humans'. *New York Times*, 25 March 2007, www.nytimes.com/2007/03/25/business/yourmoney/25Stream.html.

13   Fairwork. Fairwork Cloudwork Ratings 2023: Work in the Planetary Labour Market.

14   Gray, Mary and Suri, Siddharth. *Ghost Work: How to Stop Silicon Valley from Building a New Global Underclass*. New York: Harper Business, 2019.

15   Fairwork. Fairwork Cloudwork Ratings 2022: Work in the Planetary Labour Market.

16   Graham, Marm, Hjorth, Isis, and Lehdonvirta, Vili. 'Digital labour and development: impacts of global digital labour platforms and the gig economy on worker livelihoods'. *Transfer: European Review of Labour and Research*, Vol. 23 No. 2 (2017), pp. 135–62, doi.org/10.1177/1024258916687250.

17   Posada, Julian. 'Embedded reproduction in platform data work'. *Information, Communication & Society*, Vol. 25 No. 6 (2022), pp. 816–34, doi.org/10.1080/1369118X.2022.2049849.

18   See Schmidt, Florian. 'Crowdproduktion von trainingsdaten: Zur Rolle von Online-Arbeit beim Trainieren autonomer Fahrzeuge'. Berlin: Hans-Böckler Stiftung, 2019. https://www.econstor.eu/bitst ream/10419/201850/1/1671658035.pdf; Muldoon, James, Cant, Callum, Wu, Boxi A., Graham, Mark. 'A Typology of AI Data Work'. *Big Data & Society* (forthcoming).

19   Janah, Leila. *Give Work: Reversing Poverty One Job at a Time.* New York: Penguin, 2017.

20   Business Wire. 'Alipay Foundation and Alibaba AI Labs Launch Initiative to Bring AI-related Jobs to Women in Underdeveloped Areas in China', 6 August 2019, www.businesswire.com/news/ home/20190805005725/en/Alipay-Foundation-and-Alibaba-AI-Labs-Launch-Initiative-to-Bring-AI-related-Jobs-to-Women-in-Underdeveloped-Areas-in-China.

21   Murali, Anand. 'How India's data labellers are powering the global AI race'. *Factor Daily*, 21 March 2019, archive.factordaily.com/ indian-data-labellers-powering-the-global-ai-race/.

22   Murgia, Madhumita. 'Why computer-made data is being used to train AI models'. *Financial Times*, 19 July 2023, www.ft.com/ content/053ee253-820e-453a-a1d5-0f24985258de.

23   Linden, Alexander. 'Is Synthetic Data the Future of AI?', *Gartner*, 22 June 2022, www.gartner.com/en/newsroom/press-releases/2022-06-22-is-synthetic-data-the-future-of-ai.

24   Shumailov, Ilia, et al. 'The Curse of Recursion: Training on Generated Data Makes Models Forget', arxiv.org/abs/2305.17493.

25   Dzieza, 'AI Is a Lot of Work'.

26   Friedman, Thomas. *The World Is Flat: A Brief History of the Twenty-first Century.* New York: Farrar, Straus and Giroux, 2005.

27   Graham, Mark and Anwar, Mohammad Amir. 'The Global Gig Economy: Towards a Planetary Labour Market?'. *First Monday*, Vol. 24 No. 4 (2019), doi.org/10.5210/fm.v24i4.9913.

28   Grosfoguel, Ramón. 'The Epistemic Decolonial Turn'. *Cultural Studies*, Vol. 21 No. 2–3 (2007), pp. 211–23, doi.org/10.1080/ 09502380601162514; Mohamed, Shakir, Png, Marie-Therese and Isaac, William. 'Decolonial AI: Decolonial Theory as Sociotechnical Foresight in Artificial Intelligence'. *Philosophy & Technology*,

Vol. 33 (2020), pp. 659–84, doi.org/10.1007/s13347-020-00405-8.

29 Harvey, David. *The Condition of Postmodernity*. Oxford: Blackwell, 1989, p. 19.

2: The Engineer

1 Michie, Donald. 'Experiments on the Mechanization of Game-Learning Part I. Characterization of the Model and Its Parameters'. *The Computer Journal*, Vol. 6 No. 3 (1963), pp. 232–6, https://doi.org/10.1093/comjnl/6.3.232.

2 Narayanan, Deepak, Shoeybi, Mohammed, Casper, Jared, LeGresley, Patrick, Patwary, Mostofa, Korthikanti, Vijay, Vainbrand, Dmitri, Kashinkunti, Prethvi, Bernauer, Julie, Catanzaro, Brian, Phanishayee, Amar, Zaharia, Matei. 'Efficient Large-Scale Language Model Training on GPU Clusters Using Megatron-LM'. (2021) SC21, https://arxiv.org/abs/2104.04473.

3 For the now famous paper that introduced a new method for natural language processing and laid the groundwork for the current generation of LLMs, see Vaswani, Ashish, Shazeer, Noam, Parmar, Niki, Uszkoreit, Jakob, Jones, Llion, Gomez, Aidan N., Kaiser, Lukasz, Polosukhin, Illia. 'Attention Is All You Need'. https://arxiv.org/abs/1706.03762.

4 Bender, Emily M., Gebru, Timnit, McMillan-Major, Angelina, Mitchell, Margaret. 'On the Dangers of Stochastic Parrots: Can Language Models Be Too Big? 🦜'. *Proceedings of the 2021 ACM Conference on Fairness, Accountability, and Transparency*. FAccT '21. Association for Computing Machinery. pp. 610–23, https://dl.acm.org/doi/10.1145/3442188.3445922.

5 Common Crawl. https://commoncrawl.org/.

6 OpenAI, 'GPT-4 Technical Report'. 2023. https://cdn.openai.com/papers/gpt-4.pdf.

7 Roose, Kevin. 'A Conversation With Bing's Chatbot Left Me Deeply Unsettled'. *New York Times*, 16 February 2023, https://www.nytimes.com/2023/02/16/technology/bing-chatbot-microsoft-chatgpt.html.

8   Berglund, Lukas, Tong, Meg, Kaufmann, Max, Balesni, Mikita, Stickland, Asa Cooper, Korbak, Tomasz, Evans, Owain. 'The Reversal Curse: LLMs trained on "A is B" fail to learn "B is A"', https://arxiv.org/abs/2309.12288.

9   Bender, Emily. 'Resisting Dehumanization in the Age of "AI"', *Cognitive Dviersity 2022*. 29 July 2022, Toronto, Canada. At: https://faculty.washington.edu/ebender/papers/Bender-CogSci-2022.pdf.

10   Dennett, Daniel C. *The Intentional Stance*, Cambridge, MA: The MIT Press, 1987.

11   Goldman Sachs. 'Generative AI could raise global GDP by 7%', Goldman Sachs, 5 April 2023, https://www.goldmansachs.com/intelligence/pages/generative-ai-could-raise-global-gdp-by-7-percent.html; McKinsey Global Institute, 'Jobs lost, jobs gained: What the future of work will mean for jobs, skills, and wages'. McKinsey, 28 November 2017, https://www.mckinsey.com/featured-insights/future-of-work/jobs-lost-jobs-gained-what-the-future-of-work-will-mean-for-jobs-skills-and-wages.

12   Frey, Carl Benedikt and Osborne, Michael A. 'The Future of Employment: How Susceptible are Jobs to Computerisation?'. At: https://www.oxfordmartin.ox.ac.uk/downloads/academic/The_Future_of_Employment.pdf.

13   Gmyrek, Pawel, Berg, Janine, Bescond, David. 'Generative AI and jobs: A global analysis of potential effects on job quantity and quality'. ILO Working Paper 96. Geneva: International Labour Office, 2023. At: https://www.ilo.org/global/about-the-ilo/newsroom/news/WCMS_890740/lang--en/index.htm.

14   Meting, Chris. 'Artificial Buildup: AI Startups Were Hot In 2023, But This Year May Be Slightly Different'. CrunchBase News, 9 January 2024, https://news.crunchbase.com/ai/hot-startups-2023-openai-anthropic-forecast-2024/.

15   Eubanks, Virginia. *Automating Inequality: How High-Tech Tools Profile, Police, and Punish the Poor*. New York: St Martin's Press, 2018; O'Neil, Cathy. *Weapons of Math Destruction: How Big Data Increases Inequality and Threatens Democracy*. New York: Crown Publishing Group, 2016; Drage, Eleanor and Mackereth,

Kerry. 'Does AI Debias Recruitment? Race, Gender, and AI's "Eradication of Difference"'. *Philosophy & Technology* 35, 89 (2022).

16   For one of the first and most influential statements of this position, see Nick Bostrom's *Superintelligence: Paths, Dangers, Strategies*. Oxford: Oxford University Press, 2014.

17   See Benjamin, Ruha. *Race After Technology: Abolitionist Tools for the New Jim Code*. New York: John Wiley and Sons, 2019; Buolamwini, Joy. *Unmasking AI: My Mission to Protect What Is Human in a World of Machines*. New York: Random House, 2023; Buolamwini, Joy and Gebru, Timnit. 'Gender shades: Intersectional accuracy disparities in commercial gender classification'. *Conference on fairness, accountability and transparency*, pp. 77–91.

18   Statement on AI Risk. https://www.safe.ai/statement-on-ai-risk.

19   The 2022 Expert Survey on Progress in AI. https://aiimpacts. org/2022-expert-survey-on-progress-in-a. But note that of the 4,271 that were sent the questionnaire, only 162 responded to this question, resulting in a response rate of only 4 per cent. https:// www.warpnews.org/artificial-intelligence/no-50-of-ai-researchers-dont-believe-there-is-a-10-risk-that-ai-will-kill-us/.

20   Nature Editorial, 'Stop talking about tomorrow's AI doomsday when AI poses risks today'. *Nature*, 27 June 2023, https://www. nature.com/articles/d41586-023-02094-7.

21   Benson, Thor. 'This Disinformation Is Just for You'. *Wired*, 1 August 2023, https://www.wired.com/story/generative-ai-custom-disinformation/.

22   Robins-Early, Nick. 'Disinformation reimagined: how AI could erode democracy in the 2024 US elections'. *The Guardian*, 19 July 2023, https://www.theguardian.com/us-news/2023/jul/19/ ai-generated-disinformation-us-elections.

23   The Diary of a CEO, 'Google's DeepMind Co-founder: AI Is Becoming More Dangerous And Threatening! – Mustafa Suleyman'. YouTube, 4 September 2023, https://www.youtube.com/ watch?v=CTxnLsYHWuI.

24   Urbina, Fabio, Lentzos, Filippa, Invernizzi, Cédric, Ekins, Sean. 'Dual Use of Artificial Intelligence-powered Drug Discovery'. *Nature*

Machine Intelligence, Vol. 4 No. 3 (2022), pp. 189–91, doi:10.1038/s42256-022-00465-9.

25 According to Greek mythology, Prometheus defied the Olympian gods by stealing fire from them and giving it to humanity, representing the gift of knowledge and technology.

26 See Graham, Mark and Dittus, Martin. *Geographies of Digital Exclusion: Data and Inequality*. London: Pluto, 2022.

27 Steven T. Piantadosi (@spiantado), Tweet, 4 December 2022, https://twitter.com/spiantado/status/1599462375887114240.

28 For an overview of the literature on bias in LLMs, see Gallegos, Isabel, Rossi, Ryan, Barrow, Joe, Tanjim, Md, Kim, Sungchul, Dernoncourt, Franck, Yu, Tong, Zhang, Ruiyi, Ahmed, Nesreen. 'Bias and Fairness in Large Language Models: A Survey'. 2023. At: https://arxiv.org/pdf/2309.00770.pdf.

29 Crowell, Rachel. 'Why AI's diversity crisis matters, and how to tackle it'. *Nature*. At: https://www.nature.com/articles/d41586-023-01689-4.

30 Stanford AI Index Report 2023. https://aiindex.stanford.edu/report/.

31 West, Sarah Myer, Whittaker, Meredith and Crawford, Kate. 'Discriminating Systems: Gender, Race and Power in AI'. AI Now Institute, April 2019, https://cdn.vox-cdn.com/uploads/chorus_asset/file/16125391/discriminating_systems_041519_2.pdf.

32 2022 Taulbee Survey. https://cra.org/crn/wp-content/uploads/sites/7/2023/05/2022-Taulbee-Survey-Final.pdf.

33 Barbrook, Richard and Cameron, Andy. 'The Californian Ideology'. *Mute*, 1 September 1995, https://www.metamute.org/editorial/articles/californian-ideology.

34 Miceli, Milagros, Posada, Julian, Yang, Tianling. 'Studying Up Machine Learning Data: Why Talk About Bias When We Mean Power?'. *Proceedings of the ACM on Human–Computer Interaction*, Vol. 6 Article No. 34, pp. 1–14, doi.org/10.1145/3492853.

35 Johnson, Thadeus L., and Johnson, Natasha. 'Police Facial Recognition Technology Can't Tell Black People Apart'. *Scientific American*, 18 May 2023, https://www.scientificamerican.com/article/police-facial-recognition-technology-cant-tell-black-people-apart/; Noble, Safiya. *Algorithms of oppression: how search engines reinforce racism*. New York: NYU Press, 2018.

36   Giorno, Taylor. 'Fed watchdog warns AI, machine learning may perpetuate bias in lending'. *The Hill*, 18 July 2023, https://thehill.com/business/housing/4103358-fed-watchdog-warns-ai-machine-learning-may-perpetuate-bias-in-lending/.

37   Hao, Karen. 'AI is sending people to jail—and getting it wrong'. *MIT Technology Review*, 21 January 2019, https://www.technologyreview.com/2019/01/21/137783/algorithms-criminal-justice-ai/.

38   Burgess, Matt, Schot, Evlaine, Geiger, Gabriel. 'This Machine Could Ruin Your Life'. *Wired*, 6 March 2023, https://www.wired.com/story/welfare-algorithms-discrimination/.

39   Quijano, Anibal and Ennis, Michael. 'Coloniality of Power, Eurocentrism, and Latin America', *Nepantla: Views from South*, Vol. 1 No. 3 (2000), pp. 533–80.

40   See Nagel, Thomas. *The View From Nowhere*. Oxford: Oxford University Press, 1986.

41   On feminist standpoint theory, see Harding, Sandra (ed.), *The Feminist Standpoint Theory Reader*. New York and London: Routledge, 2004; Smith, Dorothy. 'Women's Perspective as a Radical Critique of Sociology'. *Sociological Inquiry*, Vol. 44 (1974), pp. 7–13; Hartstock, Nancy C. M. 'The Feminist Standpoint: Developing the Ground for a Specifically Feminist Historical Materialism'. In Sandra Harding and Merrill B. Hintikka (eds.), *Discovering Reality: Feminist Perspectives on Epistemology, Metaphysics, Methodology, and Philosophy of Science*. Amsterdam: Kluwer Academic Publishers, 1983. See also Collins, Patricia Hill. 'Learning from the Outsider Within: The Sociological Significance of Black Feminist Thought'. *Social Problems*, Vol. 33 No. 6 (1986), S14–S32.

42   Harding, Sandra. 'Rethinking Standpoint Epistemology: What is Strong Objectivity?'. in L. Alcoff and E. Potter (eds.), *Feminist Epistemologies*. New York and London: Routledge, 1993, pp. 49–82.

43   Cave, Stephen, Dihal, Kanta. 'The Whiteness of AI'. *Philosophy & Technology*, Vol. 33 No. 4 (2020), pp. 685–703, doi.org/10.1007/s13347-020-00415-6.

44   Crunchbase. 'Current Unicorns Tagged with AI'. https://www.crunchbase.com/lists/current-unicorns-tagged-with-ai/f296fc53-ac45-44e0-88eb-2979f7857fe2/organization.companies.

3: The Technician

1   TechTarget explains that 'Power usage effectiveness (PUE) is a metric used to determine the energy efficiency of a data center. PUE is determined by dividing the total amount of power entering a data center by the power used to run the IT equipment within it.' https://www.techtarget.com/searchdatacenter/definition/power-usage-effectiveness-PUE.

2   See Moss, Sebastian. 'Underpaid and overworked: Behind the scenes with Google's data center contractors'. *Data Center Dynamics*, 2 December 2021, https://www.datacenterdynamics.com/en/analysis/underpaid-and-overworked-behind-the-scenes-with-googles-data-center-contractors/.

3   For an analysis of Iceland's cultural history and the framing of Iceland as a 'natural home' for data centres see Johnson, Alix. 'Emplacing Data Within Imperial Histories: Imagining Iceland as Data Centers' "Natural" Home'. *Culture Machine*, Vol. 18 (2019), https://culturemachine.net/vol-18-the-nature-of-data-centers/emplacing-data/. See also Johnson, Alix. *Where Cloud is Ground: Placing Data and Making Place in Iceland*. Oakland, California: University of California Press, 2023.

4   Statista. 'Number of data centers worldwide in 2023, by country'. https://www.statista.com/statistics/1228433/data-centers-worldwide-by-country/.

5   OECD. 'OECD Economic Surveys: Iceland'. https://www.oecd.org/economy/surveys/Iceland-2021-OECD-economic-survey-overview.pdf.

6   McKinsey & Company. 'Investing in the rising data center economy'. McKinsey, 13 January 2023, https://www.mckinsey.com/industries/technology-media-and-telecommunications/our-insights/investing-in-the-rising-data-center-economy.

7   Keane, Daniel. 'More than 260 "overheating incidents" in London NHS hospitals amid climate change fears'. *The Standard*, 23 June 2023, https://www.standard.co.uk/news/health/nhs-overheating-incidents-heatwave-london-hospitals-climate-change-b1089712.html.

8   International Energy Agency. 'Data Centres and Data Transmission

Networks'. https://www.iea.org/energy-system/buildings/data-centres-and-data-transmission-networks.

9   KPMG. 'The Icelandic Data Center Industry'. March 2018, http://www.si.is/media/_eplica-uppsetning/The-Icelandic-Data-Center-Industry-FINAL.pdf.

10 Mallonee, Laura. 'Inside the Icelandic Facility Where Bitcoin Is Mined'. *Wired*, 3 November 2019, https://www.wired.com/story/iceland-bitcoin-mining-gallery/.

11 Bjarnason, Egill. 'Iceland is a bitcoin miner's haven, but not everyone is happy'. Al Jazeera, 15 April 2019, https://www.aljazeera.com/features/2019/4/15/iceland-is-a-bitcoin-miners-haven-but-not-everyone-is-happy.

12 Le Page, Michael. 'The green tech that could help Iceland become carbon neutral by 2040'. *New Scientist*, 4 January 2023, https://www.newscientist.com/article/mg25634202-900-the-green-tech-that-could-help-iceland-become-carbon-neutral-by-2040/.

13 This is aside from energy exported indirectly through bars of aluminium produced in Iceland's large aluminimum smelters, which generated $3 billion in 2022. CEIC, 'Iceland Aluminum: Exports', https://www.ceicdata.com/en/indicator/iceland/aluminum-exports.

14 See Zook, Matthew and Grote, Michael H. 'The microgeographies of global finance: High-frequency trading and the construction of information inequality'. *Environment & Planning A*, Vol. 49 No. 1 (2017), pp. 121–40, doi/10.1177/0308518X16667298.

15 Verizon, 'IP Latency Statistics'. https://www.verizon.com/business/en-gb/terms/latency/.

16 Starosielski, Nicole. *The Undersea Network*. Durham, North Carolina: Duke University Press, 2015.

17 Tarnoff, Ben. *Internet for the People: The Fight for Our Digital Future*. London: Verso, 2022.

18 Starosielski, *The Undersea Network*.

19 Burrington, Ingrid. 'How Railroad History Shaped Internet History'. *The Atlantic*, 24 November 2015, https://www.theatlantic.com/technology/archive/2015/11/how-railroad-history-shaped-internet-history/417414/.

20 See Kennedy, P. M. 'Imperial Cable Communications and Strategy,

1870–1914'. *The English Historical Review*, Vol. 86 No. 341 (1971), pp. 728–52.

21   Burns, Bill. 'History of the Atlantic Cable & Undersea Communications', https://atlantic-cable.com/Maps/index.htm.

22   Kennedy, 'Imperial Cable Communications and Strategy, 1870–1914'.

23   Haigh, Kenneth Richardson. *Cable Ships and Submarine Cables*. London: Adlard Coles, 1968.

24   Headrick, D. R. and Griset, P. 'Submarine Telegraph Cables: Business and Politics, 1838–1939'. *The Business History Review*, Vol. 75 No. 3 (2001), pp. 543–578.

25   Corera, Gordon. 'How Britain pioneered cable-cutting in World War One'. BBC News, 17 September 2017, https://www.bbc.co.uk/news/world-europe-42367551.

26   Parfitt, Tom. 'Georgian woman cuts off web access to whole of Armenia'. *The Guardian*, 6 April 2011, https://www.theguardian.com/world/2011/apr/06/georgian-woman-cuts-web-access.

27   Starosielski, *The Undersea Network*.

28   Starosielski, *The Undersea Network*.

29   Cochrane, Paul. '"Digital Suez": How the internet flows through Egypt – and why Google could change the Middle East'. *Middle East Eye*, 3 March 2021, https://www.middleeasteye.net/news/google-egypt-suez-digital-internet-flow-change-middle-east.

30   Data Center Dynamics. 'NEC to build world's highest capacity submarine cable for Facebook, shuttling 500Tbps from US to Europe'. *Data Center Dynamics*, 12 October 2021, https://www.datacenterdynamics.com/en/news/nec-to-build-worlds-highest-capacity-submarine-cable-for-facebook-shuttling-500tbps-from-us-to-europe/.

31   Heller, Martin. 'Large language models: The foundations of generative AI'. *InfoWorld*, 14 November 2023, https://www.infoworld.com/article/3709489/large-language-models-the-foundations-of-generative-ai.html.

32   Appenzeller, Guido, Bornstein, Matt and Casado, Martin. 'Navigating the High Cost of AI Compute'. Andreessen Horowitz, 27 April 2023, https://a16z.com/navigating-the-high-cost-of-ai-compute/.

33   McKinsey & Company. 'Investing in the rising data center economy'.

34   Statista. 'Share of worldwide hyperscale data center capacity in 2nd quarter 2022, by region', https://www.statista.com/statistics/1350992/global-hyperscale-data-center-capacity/.

35   JLL. '2023 Global Data Center Outlook'. 13 April 2023, https://www.us.jll.com/en/trends-and-insights/research/data-center-outlook.

36   Synergy Research Group. 'Hyperscale Data Center Capacity to Almost Triple in Next Six Years, Driven by AI'. 17 October 2023, https://www.srgresearch.com/articles/hyperscale-data-center-capacity-to-almost-triple-in-next-six-years-driven-by-ai.

37   Kidd, David. 'The Data Center Capital of the World Is in Virginia'. *Governing*, 27 July 2023, https://www.governing.com/infrastructure/the-data-center-capital-of-the-world-is-in-virginia.

38   Synergy Research Group. 'Microsoft, Amazon and Google Account for Over Half of Today's 600 Hyperscale Data Centers'. 26 January 2021, https://www.srgresearch.com/articles/microsoft-amazon-and-google-account-for-over-half-of-todays-600-hyperscale-data-centers.

39   Kim, Tae. 'Meta Boosts Its Spending Plans for 2024. It May Be Good News for Nvidia'. Barron's, 2 February 2024, https://www.barrons.com/livecoverage/apple-amazon-meta-facebook-earnings-stock-price-today/card/meta-boosts-its-spending-plans-for-2024-it-may-be-good-news-for-nvidia--dRwAbF7R9LGAWzaOYvVo.

40   Field, Hayden. 'Tech execs are telling investors they have to spend money to make money on AI'. CNBC, 2 February 2024, https://www.cnbc.com/2024/02/02/techs-new-ai-game-plan-spend-money-to-make-money.html.

41   Vipra, Jai and West, Sarah Myers. 'Computational Power and AI'. AI Now Institute, 27 September 2023, https://ainowinstitute.org/publication/policy/compute-and-ai.

42   Cao, Sissi. 'Amid a Heated A.I. Race, Apple Struggles to Retain Top Talent'. *Observer*, 2 May 2023, https://observer.com/2023/05/apple-ai-chief-chatgpt-google/.

43   Frier, Sarah. 'Meta, Google Talent Leaving for AI Startups, Khosla Says'. Bloomberg, 4 May 2023, https://www.bloomberg.com/news/newsletters/2023-05-04/meta-google-talent-leaving-for-ai-startups-khosla-says.

44   Hurd, Tom. 'The State of AI Talent 2024'. Report published by
     Zeki. https://www.thezeki.com/the-state-of-ai-talent-2024.

45   EirGrid and Soni. 'All-Island Generation Capacity Statement
     2020–2029'. https://www.eirgridgroup.com/site-files/library/EirGrid/
     All-Island-Generation-Capacity-Statement-2020-2029.pdf.

46   Not Here Not Anywhere: For A Fossil Free Future. https://
     notherenotanywhere.com/campaigns/data-centres/.

47   Swinhoe, Dan. 'EirGrid says no new applications for data centers in
     Dublin until 2028'. *Data Center Dynamics*, 11 January 2022,
     https://www.datacenterdynamics.com/en/news/eirgrid-says-no-new-
     applications-for-data-centers-in-dublin-till-2028/.

48   Australian Energy Council. 'Data centres: A 24hr power source?'.
     12 October 2023, https://www.energycouncil.com.au/analysis/data-
     centres-a-24hr-power-source.

49   Rogoway, Mike. 'Google's water use is soaring in The Dalles,
     records show, with two more data centers to come'. *The
     Oregonian*, 17 December 2022, https://www.oregonlive.com/silicon-
     forest/2022/12/googles-water-use-is-soaring-in-the-dalles-records-
     show-with-two-more-data-centers-to-come.html.

50   Livingstone, Grace. '"It's pillage": thirsty Uruguayans decry
     Google's plan to exploit water supply'. *The Guardian*, 11 July
     2023, https://www.theguardian.com/world/2023/jul/11/uruguay-
     drought-water-google-data-center.

51   Google Sustainability. '2023 Environmental Report'. https://
     sustainability.google/reports/google-2023-environmental-report/.

52   Moss, Sebastian. 'Underpaid and overworked: Behind the scenes
     with Google's data center contractors'.

53   Stanford AI Index Report 2023. https://aiindex.stanford.edu/
     wp-content/uploads/2023/04/HAI_AI-Index-Report_2023.pdf.

54   National Committee on U.S.–China Relations. 'Kai-Fu Lee:
     AI Superpowers'. https://www.youtube.com/watch?v=0-
     8VccRDHgY.

55   International Energy Agency. 'Critical Minerals Market Review 2023'.
     https://www.iea.org/reports/critical-minerals-market-review-2023.

56   ING. 'China strikes back in the tech war, restricting exports of
     gallium and germanium', 7 July 2023, https://www.ing.com/

Newsroom/News/China-strikes-back-in-the-tech-war-restricting-exports-of-gallium-and-germanium.htm.

57   International Energy Agency. 'Critical Minerals Market Review 2023'.

58   Stanford AI Index Report 2023.

59   Leswing, Kif. 'Meet the $10,000 Nvidia chip powering the race for A.I.'. CNBC, 23 Februrary 2023, https://www.cnbc.com/2023/02/23/nvidias-a100-is-the-10000-chip-powering-the-race-for-ai-.html.

60   Khan, Saif M., and Mann, Alexander. 'AI Chips: What They Are and Why They Matter'. Center for Security and Emerging Technology, April 2020, http://cset.georgetown.edu/research/ai-chips-what-they-are-and-why-they-matter/.

61   Toews, Rob. 'The Geopolitics Of AI Chips Will Define The Future Of AI'. *Forbes*, 7 May 2023, https://www.forbes.com/sites/robtoews/2023/05/07/the-geopolitics-of-ai-chips-will-define-the-future-of-ai/.

62   Mann, Tobias. 'TSMC warns AI chip crunch will last another 18 months'. *The Register*, 8 September 2023, https://www.theregister.com/2023/09/08/tsmc_ai_chip_crunch/.

63   Financial Times. 'Saudi Arabia and UAE race to buy Nvidia chips to power AI ambitions'. *Financial Times*, 14 August 2023, https://www.ft.com/content/c93d2a76-16f3-4585-af61-86667c5090ba.

64   Gaida, Jamie, Wong-Leung, Jenny, Robin, Stephen, Dave, Pilgrim, Danielle. 'ASPI's Critical Technology Tracker – Sensors & Biotech updates'. Australian Strategic Policy Institute, https://www.aspi.org.au/report/critical-technology-tracker.

65   The Select Committee on the CCP. 'Leveling the Playing Field: How to Counter the Chinese Communist Party's Economic Aggression'. Hearing at Longworth House Office Building, Room 1310, Washington D.C., 17 May 2023, https://selectcommitteeontheccp.house.gov/committee-activity/hearings/hearing-notice-leveling-playing-field-how-counter-chinese-communist.

66   New York Times. 'Pressured by Biden, A.I. Companies Agree to Guardrails on New Tools'. *New York Times*, 21 July 2023, https://www.nytimes.com/2023/07/21/us/politics/ai-regulation-biden.html.

67   OpenAI, 'Frontier Model Forum'. https://openai.com/blog/frontier-model-forum.

4: The Artist

1   Yeo, Amanda. 'Netflix is getting blasted for using AI art in an anime instead of hiring artists'. Mashable, 2 February 2023, https://mashable.com/article/netflix-ai-art-anime-boy-dog.

2   Creamer, Ella. 'Authors file a lawsuit against OpenAI for unlawfully "ingesting" their books'. *The Guardian*, 5 July 2023, https://www.theguardian.com/books/2023/jul/05/authors-file-a-lawsuit-against-openai-for-unlawfully-ingesting-their-books.

3   LAION-5B. https://laion.ai/blog/laion-5b/.

4   Vincent, James. 'AI art tools Stable Diffusion and Midjourney targeted with copyright lawsuit'. The Verge, 16 January 2023, https://www.theverge.com/2023/1/16/23557098/generative-ai-art-copyright-legal-lawsuit-stable-diffusion-midjourney-deviantart. As of December 2023, the applicants have ammended their original lawsuit and added a new defendant, Runway AI, and included more details of the alleged infringement.

5   Salkowitz, Rob. 'Midjourney founder David Holz on the impact of AI on art imagination and the creative economy'. *Forbes*, 16 September 2023, https://www.forbes.com/sites/robsalkowitz/2022/09/16/midjourney-founder-david-holz-on-the-impact-of-ai-on-art-imagination-and-the-creative-economy/.

6   Heikkilä, Melissa. 'This new data poisoning tool lets artists fight back against generative AI'. *MIT Technology Review*, 23 October 2023, https://www.technologyreview.com/2023/10/23/1082189/data-poisoning-artists-fight-generative-ai/.

7   See Rothman, Jennifer. *The Right of Publicity: Privacy Reimagined for a Public World*. Cambridge, Cambridge, MA: Harvard University Press, 2018.

8   Obedkov, Evgeny. 'What game devs think about generative AI: "World and mission design are about as AI solvable as neurosurgery"'. *Game World Observer*, 14 December 2022, https://gameworldobserver.com/2022/12/14/generative-ai-game-developers-opinion-cyberpunk-2077-rdr2.

9   The Creative Independent. 'A study on the financial state of visual artists today'. thecreativeindependent.com/artist-survey/.

10   Bedingfield, Will. 'Hollywood Writers Reached an AI Deal That Will Rewrite History'. *Wired*, 22 September 2023, https://www.wired.co.uk/article/us-writers-strike-ai-provisions-precedents.

11   SAG-AFTRA. 'A Message from the SAG-AFTRA President and Chief Negotiator'. 13 July 2023, https://www.sagaftra.org/message-sag-aftra-president-and-chief-negotiator.

12   The Alliance of Motion Picture and Television Producers disputed this claim. Chmielewski, Dawn. 'Black Mirror: Actors and Hollywood battle over AI digital doubles'. Reuters, 14 July 2023, https://www.reuters.com/business/media-telecom/union-fears-hollywood-actors-digital-doubles-could-live-for-one-days-pay-2023-07-13/.

13   At the time of writing, only a draft of the agreement had been released. https://www.sagaftra.org/files/2023%20SAG-AFTRA%20TV-Theatrical%20MOA_F.pdf.

14   Weiss, Laura. 'SAG-AFTRA's new contract falls short on protections from Artificial Intelligence'. *Prism*, 5 December 2023, https://prismreports.org/2023/12/05/sag-aftra-contract-falls-short-ai-protections/.

15   United Voice Artists. https://www.unitedvoiceartists.com/.

16   United Voice Artists. 'World Voice Professionals Speaking Up'. https://www.unitedvoiceartists.com/wp-content/uploads/2023/05/240522_UVA_Manifesto_Validated_8.pdf.

17   Quoted in Hartree, Douglas. *Calculating instruments and machines*. Urbana-Champaign: University of Illinois Press, 1949, p. 70.

18   Turing, Alan. 'Computing machinery and intelligence', *Mind 59* (1950) pp. 433–60.

19   Press Association. 'Computer simulating 13-year-old boy becomes first to pass Turing test'. *The Guardian*, 9 June 2014, https://www.theguardian.com/technology/2014/jun/08/super-computer-simulates-13-year-old-boy-passes-turing-test.

20   Bringsjord, Selmer, Bello, Paul and Ferrucci, David. 'Creativity, the Turing Test, and the (Better) Lovelace Test'. *Minds and Machines* 11, pp. 3–27 (2001), https://doi.org/10.1023/A:1011206622741.

21   On debates within computational creativity on different evaluation methodologies, see Jordanous, Anna. 'Evaluating evaluation: Assessing progress in computational creativity research'. In:

Proceedings of the Second International Conference on Computational Creativity (ICCC-11) (2011), pp. 102–107.

22 Veale, Tony, and Cardoso, F. Amílcar (eds.). *Computational Creativity: The Philosophy and Engineering of Autonomously Creative Systems*. Cham: Springer, 2019.

23 Quoted in Marks, Robert J. *Non-Computable You: What You Do that Artificial Intelligence Never Will*. Seattle: Discovery Institute Press, 2022.

24 Quoted in Du Sautoy, Marcus. *The Creativity Code: How AI is learning to write, paint and think*. London: Fourth Estate, 2020.

25 Silver, D., Hubert, T., Schrittwieser, J., Antonoglou, I., Lai, M., Guez, A., Lanctot, M., Sifre, L., Kumaran, D., Graepel, T., Lillicrap, T., Simonyan, K., and Hassabis, D. 'Mastering Chess and Shogi by Self-Play with a General Reinforcement Learning Algorithm' (2017). ArXiv. /abs/1712.01815.

26 Cone, Gabe. 'AI Art at Christie's Sells for $432,500'. *New York Times*, 25 October 2018, https://www.nytimes.com/2018/10/25/arts/design/ai-art-sold-christies.html.

27 More recently, image generators that use GANs have become less popular in favour of those that use a diffusion technique. These diffusion models operate by learning the basic structure of images and then applying 'noise' to these images in a process that imitates the diffusion of fluid dynamics. This allows them to recover images by gradually adding detail in progressive layers. See Song, Jiaming, Meng, Chenlin, and Ermon, Stefano. 'Denoising Diffusion Implicit Models' (2020), https://doi.org/10.48550/ARXIV.2010.02502.

28 Marks, Robert J. and Bringsjord, Selmer. 'Thinking Machines? Has the Lovelace Test Been Passed?'. *Mind Matters*, 17 April 2020, 'https://mindmatters.ai/2020/04/thinking-machines-has-the-lovelace-test-been-passed/.

29 Schaub, Michael. 'Is the future award-winning novelist a writing robot?'. *Los Angeles Times*, 22 March 2016, https://www.latimes.com/books/jacketcopy/la-et-jc-novel-computer-writing-japan-20160322-story.html.

30 Merchant, Brian. 'When an AI Goes Full Jack Kerouac'. *The*

Atlantic, 1 October 2018, https://www.theatlantic.com/technology/archive/2018/10/automated-on-the-road/571345/.

31 Tapper, James. 'Authors shocked to find AI ripoffs of their books being sold on Amazon'. *The Guardian*, 30 September 2023, https://www.theguardian.com/technology/2023/sep/30/authors-shocked-to-find-ai-ripoffs-of-their-books-being-sold-on-amazon.

32 Jiang, Harry, Lauren Brown, Jessica Cheng, Anonymous Artist, Mehtab Khan, Abhishek Gupta, Deja Workman, Alex Hanna, Jonathan Flowers, and Timnit Gebru. 'AI Art and its Impact on Artists'. In: AAAI/ACM Conference on AI, Ethics, and Society. 8–10 August 2023, https://doi.org/10.1145/3600211.3604681.

33 Cave, Nick. The Red Hand Files No. 218, January 2023, https://www.theredhandfiles.com/chat-gpt-what-do-you-think/.

34 Eliot, George. 'The Natural History of German Life', in *George Eliot: Selected Critical Writings*, ed. by Rosemary Ashton. Oxford: Oxford University Press, 1992.

35 Kant, Immanuel. *The Critique of the Power of Judgement*. Edited and translated by Paul Guyer. Cambridge: Cambridge University Press, 2013.

36 Kant, *The Critique of the Power of Judgement*, §44, 5: 306.

37 Heidegger, Martin. 'The Origin of the Work of Art'. In *Poetry, Language, Thought*. A. Hofstadter (trans.). New York: Harper & Row, 1971.

38 Orwell, George. 'Why I Write'. *Gangrel* No. 4, Summer 1946, available at: https://www.orwellfoundation.com/the-orwell-foundation/orwell/essays-and-other-works/why-i-write/.

39 Benjamin, Walter. 'The Work of Art in the Age of Mechanical Reproduction'. In *Illuminations*, edited by Hannah Arendt. New York: Schocken Books, 1969.

40 Benjamin, 'The Work of Art in the Age of Mechanical Reproduction'.

41 Bellinetti, Caterina. '"From Today Painting is Dead": Photography's Revolutionary Effect'. *Art & Object*, 8 April 2019, https://www.artandobject.com/news/today-painting-dead-photographys-revolutionary-effect.

42 Baudelaire, Charles. 'On Photography', from *The Salon of 1859*.

https://www.csus.edu/indiv/o/obriene/art109/readings/11%20baudelaire
%20photography.htm.

43   Victoria & Albert Museum. 'Julia Margaret Cameron's working
     methods'. https://www.vam.ac.uk/articles/julia-margaret-camerons-
     working-methods.

44   Silva, Eva. 'How Photography Pioneered a New Understanding of
     Art'. The Collector, 4 June 2022, https://www.thecollector.com/
     how-photography-transformed-art/.

45   Anna Ridler's Personal Website. https://annaridler.com/.

46   'Selected AI works by Helena Sarin'. This is Paper. https://www.
     thisispaper.com/mag/selected-ai-nft-works-helena-sarin.

47   'AI Has Already Created As Many Images As Photographers Have
     Taken in 150 Years. Statistics for 2023'. Everypixel Journal. https://
     journal.everypixel.com/ai-image-statistics.

48   Kelly, Kevin. 'Picture Limitless Creativity at Your Fingertips'.
     Wired, 17 November 2022, https://www.wired.com/story/picture-
     limitless-creativity-ai-image-generators/.

5: The Operator

1   Donnelly, Tom, Begley, Jason, Collis, Clive. 'The West Midlands
     automotive industry: the road downhill'. Business History, Vol. 59
     No. 1 (2017), pp. 56–74.

2   Healey, Mick J., and Clark, David. 'Industrial Decline in a Local
     Economy: The Case of Coventry, 1974–1982'. Environment
     and Planning A: Economy and Space, Vol. 17 No. 10 (1985),
     pp. 1351–67.

3   Healey, Mick J., and Clark, David. 'Industrial Decline in a Local
     Economy: The Case of Coventry, 1974–1982'.

4   O'Neill, Sean. 'Solving some of the largest, most complex
     operations problems'. Amazon Science, 14 October 2022, https://
     www.amazon.science/latest-news/solving-some-of-the-largest-most-
     complex-operations-problems; Staff Writer. 'How peak events like
     Prime Day helped Amazon navigate the pandemic'. Amazon
     Science, 11 July 2022, https://www.amazon.science/latest-news/
     solving-some-of-the-largest-most-complex-operations-problems.

5   O'Neill, 'Solving some of the largest, most complex operations problems'.

6   Staff Writer, 'How peak events like Prime Day helped Amazon navigate the pandemic'.

7   Amazon, 'The Science behind the New FBA Capacity Management System'. Amazon Science, 2023, https://www.amazon.science/news-and-features/science-behind-fulfillment-by-amazon-fba-capacity-management-system.

8   Amazon Jobs, 'Supply Chain Optimization Technologies'. 'https://www.amazon.jobs/en/teams/scot.

9   Amazon. 'Maximizing the Efficiency of Amazon's Own Delivery Networks'. Amazon Science, 14 October 2022, https://www.amazon.science/blog/maximizing-the-efficiency-of-amazons-own-delivery-networks.

10   Lecher, Colin. 'How Amazon Automatically Tracks and Fires Warehouse Workers for "Productivity"'. The Verge, 25 April 2019, https://www.theverge.com/2019/4/25/18516004/amazon-warehouse-fulfillment-centers-productivity-firing-terminations.

11   Bezos, Jeff. '2020 Letter to Shareholders'. Amazon, 15 April 2021, https://www.aboutamazon.com/news/company-news/2020-letter-to-shareholders.

12   McIntyre, Nimah, and Bradbury, Rosie. 'The Eyes of Amazon: A Hidden Workforce Driving a Vast Surveillance System'. The Bureau of Investigative Journalism, 2022, https://www.thebureauinvestigates.com/stories/2022-11-21/the-eyes-of-amazon-a-hidden-workforce-driving-a-vast-surveillance-system.

13   McIntyre and Bradbury, 'The Eyes of Amazon: A Hidden Workforce Driving a Vast Surveillance System'.

14   Rey, Jason Del, and Ghaffray, Shirin. 'Leaked: Confidential Amazon Memo Reveals New Software to Track Unions'. Vox, 6 October 2020, https://www.vox.com/recode/2020/10/6/21502639/amazon-union-busting-tracking-memo-spoc.

15   Weatherbed, Jess. 'Amazon's in-van surveillance footage of delivery drivers is leaking online'. The Verge, 18 July 2023, https://www.theverge.com/2023/7/18/23798611/amazon-van-driver-surveillance-camera-footage-leak-reddit.

16   McCarthy, John, Minsky, Marvin L., Rochester, Nathaniel, Shannon, Claude E. 'A Proposal for the Dartmouth Summer Research Project on Artificial Intelligence'. 31 August 1955, reprinted in *AI Magazine*, Vol. 27 No. 4 (2006).

17   Quoted in Beynon, Huw. *Working for Ford*. New York: Allen Lane, 1976.

18   Nye, David E. *America's Assembly Line*. Cambridge, Mass: The MIT Press, 2013.

19   Benyon, *Working for Ford*.

20   Panzieri, Raniero. 'The Capitalist Use of Machinery: Marx versus the "Objectivists"'. In: *Outlines of a Critique of Technology*, Phil Slater (ed.), pp. 44–68. London: Atlantic Highlands, 1980; Noble, David F. *Forces of Production: A Social History of Industrial Automation*. New Brunswick, NJ: Transaction Publishers, 2011; Mueller, Gavin. *Breaking Things at Work: The Luddites Are Right about Why You Hate Your Job*. London: Verso, 2021.

21   Benanav, Aaron. *Automation and the Future of Work*. London: Verso, 2020.

22   See Braverman, Harry. *Labour and Monopoly Capital: The Degradation of Work in the Twentieth Century*. New York: Monthly Review Press, 1998.

23   Del Rey, Jason. 'Leaked Amazon memo warns the company is running out of people to hire'. *Vox*, 17 June 2022, https://www.vox.com/recode/23170900/leaked-amazon-memo-warehouses-hiring-shortage.

24   Segal, Edward. 'Amazon Responds To Release Of Leaked Documents Showing 150% Annual Employee Turnover'. *Forbes*, 24 October 2022, 'https://www.forbes.com/sites/edwardsegal/2022/10/24/amazon-responds-to-release-of-leaked-documents-showing-150-annual-employee-turnover/?sh=4a90f9051d0b.

25   Kelly, Jack. 'A Hard-Hitting Investigative Report Into Amazon Shows That Workers' Needs Were Neglected In Favor Of Getting Goods Delivered Quickly'. *Forbes*, 25 October 2021, https://www.forbes.com/sites/jackkelly/2021/10/25/a-hard-hitting-investigative-report-into-amazon-shows-that-workers-needs-were-neglected-in-favor-of-getting-goods-delivered-quickly/.

26   See, for example, the pioneering work in Moore, Phoebe, Robinson, Andrew. 'The quantified self: What counts in the neoliberal workplace'. *New Media & Society*, Vol. 18 No. 11 (2016), pp. 2,774–92; Moore, Phoebe. 'Tracking affective labour for agility in the quantified workplace'. *Body & Society*, Vol. 24 No. 3 (2018), pp. 39–67.

27   Microsoft. '2022 Work Trend Index Survey'. 22 September 2022, https://www.microsoft.com/en-us/worklab/work-trend-index/hybrid-work-is-just-work.

28   Turner, Jordan. 'The Right Way to Monitor Your Employee Productivity'. *Gartner*, 9 June 2022, https://www.gartner.com/en/articles/the-right-way-to-monitor-your-employee-productivity.

29   Kantor, Jodi, and Sundaram, Arya. 'The Rise of the Worker Productivity Score'. *New York Times*, 14 August 2022, https://www.nytimes.com/interactive/2022/08/14/business/worker-productivity-tracking.html.

30   Cutter, Chip, and Chen, Te-Ping. 'Bosses Aren't Just Tracking When You Show Up to the Office but How Long You Stay'. *The Wall Street Journal*, 25 September 2023, https://www.wsj.com/lifestyle/careers/attention-office-resisters-the-boss-is-counting-badge-swipes-5fa37ff7.

31   BBC News. 'Court win for man fired for not keeping webcam on'. BBC News, 11 October 2022, https://www.bbc.co.uk/news/technology-63203945.

32   Nguyen, Aiha. 'The Constant Boss'. *Data & Society*, 19 May 2021, https://datasociety.net/library/the-constant-boss/.

33   Negrón, Wilneida. 'Little Tech is Coming for Workers'. Coworker, 2021, https://home.coworker.org/wp-content/uploads/2021/11/Little-Tech-Is-Coming-for-Workers.pdf.

34   Temperton, James. 'Uber is tracking dangerous drivers with smartphone sensors'. *Wired*, 26 January 2016, https://www.wired.co.uk/article/uber-dangerous-drivers-smartphone-tracking.

35   Migliano, Simon. 'Employee Monitoring Software Demand up 60% since 2019'. Top10VPN, 8 August 2023, https://www.top10vpn.com/research/employee-monitoring-software-privacy/.

36   Teale, Chris. 'Many Tech Employees Say They'd Quit Rather Than

Be Monitored During Work'. *Morning Consult*, 31 May 2022, https://pro.morningconsult.com/trend-setters/tech-workers-survey-surveillance.

37   Fennell, Andrew. 'Employee Monitoring Statistics'. StandOutCV, 2023, https://standout-cv.com/employee-monitoring-study.

38   Carnegie, Megan. 'The Creepy Rise of Bossware'. *Wired*, 23 July 2023, https://www.wired.co.uk/article/creepy-rise-bossware.

39   Patel, Vishal, Chesmore, Austin, Legner, Christopher M., Pandey, Santosh. 'Trends in Workplace Wearable Technologies and Connected-Worker Solutions for Next-Generation Occupational Safety, Health, and Productivity'. *Advanced Intelligent Systems*, Vol. 4 No. 1 (2022), doi.org/10.1002/aisy.202100099.

40   Stanford Social Innovation Review. 'Bossware Is Coming for You: Worker Surveillance Technology Is Everywhere'. *Stanford Social Innovation Review*, https://ssir.org/videos/entry/bossware_is_coming_for_you_worker_surveillance_technology_is_everywhere.

41   Perceptyx. 'Sense: Uncover valuable insights from moments that matter across the employee lifecycle'. https://go.perceptyx.com/platform/sense-employee-lifecycle-surveys.

42   Smith, Genevieve, and Rustagi, Ishita. 'Workplace AI Wants to Help You Belong'. *Stanford Social Innovation Review*, 14 September 2022, https://ssir.org/articles/entry/workplace_ai_wants_to_help_you_belong.

43   Ruby, Daniel. '41 AI Recruitment Statistics 2023 (Facts & Hiring Trends)'. DemandSage, 6 July 2023, https://www.demandsage.com/ai-recruitment-statistics/.

44   Barrett, Jonathan, and Convery, Stephanie. 'Robot recruiters: can bias be banished from AI hiring?'. *The Guardian*, 26 March 2023, https://www.theguardian.com/technology/2023/mar/27/robot-recruiters-can-bias-be-banished-from-ai-recruitment-hiring-artificial-intelligence; Ayoub, Sarah. 'Recruitment by robot: how AI is changing the way Australians get jobs'. *The Guardian*, 22 October 2023, https://www.theguardian.com/technology/2023/oct/23/ai-recruitment-job-search-artificial-intelligence-employment.

45   Weissmann, Jordan. 'Amazon Created a Hiring Tool Using A.I. It Immediately Started Discriminating Against Women'. *Slate*, 10

October 2018, https://slate.com/business/2018/10/amazon-artificial-intelligence-hiring-discrimination-women.html.

46   Dastin, Jeffery. 'Insight – Amazon scraps secret AI recruiting tool that showed bias against women'. Reuters, 28 October 2018, https://www.reuters.com/article/idUSKCN1MK0AG/.

47   Drage, Eleanor, and Mackereth, Kerry. 'Does AI Debias Recruitment? Race, Gender, and AI's "Eradication of Difference"'. *Philosophy & Technology* 35 (2022), doi.org/10.1007/s13347-022-00543-1.

48   Fergus, J., 'A bookshelf in your job screening video makes you more hirable to AI'. *Input*, 18 February 2021, https://www.inverse.com/input/culture/a-bookshelf-in-your-job-screening-video-makes-you-more-hirable-to-ai; Schellmann, Hilke, and Wall, Sheridan. 'We tested AI interview tools. Here's what we found'. *MIT Technology Review*, 7 July 2021, https://www.technologyreview.com/2021/07/07/1027916/we-tested-ai-interview-tools.

49   ModernHire. 'Automated Interview Creator'. https://modernhire.com/platform/automated-interview-creator/.

50   Barnes, Patricia. 'EPIC Asks Federal Trade Commission To Regulate Use Of Artificial Intelligence In Pre-Employment Screenings'. *Forbes*, 3 February 2022, https://www.forbes.com/sites/patriciagbarnes/2020/02/03/group-asks-federal-trade-commission-to-regulate-use-of-artificial-intelligence-in-pre-employment-screenings/.

51   Ryan-Mosley, Tate. 'Why everyone is mad about New York's AI hiring law'. *MIT Technology Review*, 10 July 2023, https://www.technologyreview.com/2023/07/10/1076013/new-york-ai-hiring-law/.

52   European Comission. 'Regulatory framework proposal on artificial intelligence'. https://digital-strategy.ec.europa.eu/en/policies/regulatory-framework-ai.

53   Ghaffary, Shirin, and Del Rey, Jason. 'Amazon employees fear HR is targeting minority and activism groups in email monitoring program'. *Vox*, 24 September 2020, https://www.vox.com/recode/2020/9/24/21455196/amazon-employees-listservs-minorities-underrepresented-groups-worker-dissent-unionization.

54   Vallas, Steven P., Hannah Johnston, and Mommadova, Yana. 'Prime

Suspect: Mechanisms of Labor Control at Amazon's Warehouses'. *Work and Occupations*, Vol. 49 No. 4 (2022), pp. 421–56, doi. org/10.1177/07308884221106922.

55   See, for example, the debate over call centres and worker resistance: Bain, Peter, and Phil Taylor. 'Entrapped by the "Electronic Panopticon"? Worker Resistance in the Call Centre'. *New Technology, Work and Employment*, Vol. 15 No. 1 (2000), pp. 2–18, doi.org/10.1111/1468-005X.00061; Woodcock, Jamie. *Working the Phones*. London: Pluto Press, 2017; Ferrari, Fabian, and Graham, Mark. 'Fissures in algorithmic power: platforms, code, and contestation'. *Cultural Studies* (2021), doi.org/10.1080/09 502386.2021.1895250.

56   Silver, Beverly. *Forces of Labour: Workers' Movements and Globalisation since 1870*. Cambirdge: Cambridge University Press, 2003.

57   Fine, Sidney. *Sit-down: The General Motors Strike of 1936–1937*. Ann Arbor: University of Michigan Press, 1959; Rawick, George. 'Notes on the American Working Class'. *Speak Out*, 1968. https:// www.marxists.org/archive/rawick/1968/06/us-working-class.htm.

58   Cant, Callum. *Riding for Deliveroo: Resistance in the New Economy*. Cambridge: Polity Press, 2019.

59   Delfanti, Alessandro. *The Warehouse: Workers and Robots at Amazon*. London: Pluto Press, 2021.

60   Chua, Charmaine, and Cox, Spencer. 'Battling the Behemoth: Amazon and the Rise of America's New Working Class'. *Socialist Register* (2023), https://socialistregister.com/index.php/srv/article/ view/39597; Moody, Kim. *On New Terrain: How Capital Is Reshaping the Battleground of Class War*. Chicago: Haymarket Books, 2017.

61   Anon. 'Wildcat Strike at Amazon'. *Notes From Below*, 2022, https://notesfrombelow.org/article/wildcat-strike-amazon.

62   Anon. 'How the Amazon Wildcat Spread'. *Notes From Below*, 2022, https://notesfrombelow.org/article/how-amazon-wildcat-spread.

63   Childs, Simon. 'Amazon Hit by Strikes Across the Globe'. Novara Media, 2022, https://novaramedia.com/2022/08/19/amazon-hit-by-strikes-across-the-globe/.

64   Boewe, Jorn, and Schulten, Johannes. 'The Long Struggle of the Amazon Employees'. Berlin: Rosa-Luxemburg-Stiftung, 2019. https://www.rosalux.de/en/publication/id/8529/the-long-struggle-of-the-amazon-employees; Cant, Callum. 'Mapping the Amazon Strikes'. *Notes From Below*, 2022, https://notesfrombelow.org/article/mapping-amazon-strikes.

65   Boewe, Jorn, and Johannes Schulten. 'The Long Struggle of the Amazon Employees'. Berlin: Rosa-Luxemburg-Stiftung, 2019. https://www.rosalux.de/en/publication/id/8529/the-long-struggle-of-the-amazon-employees.

66   Weise, Karen, and Scheiber, Noam. 'Amazon Labor Union Loses Election at Warehouse Near Albany'. *New York Times*, 18 October 2022, https://www.nytimes.com/2022/10/18/technology/amazon-labor-union-alb1.html.

67   Gall, Gregor. 'Union Busting at Amazon.Com in Britain'. Indymedia, 2004, https://www.indymedia.org.uk/en/2004/01/284179.html.

6: The Investor

1   S&P 500's information technology sector, which includes large tech companies, increased 43.89% on the year in 2020. See Scehid, Brian. 'Driven by big tech's pandemic gains, S&P 500's 2020 surge masks uneven recovery'. *S&P Global Market Intelligence*, 4 January 2021, https://www.spglobal.com/marketintelligence/en/news-insights/latest-news-headlines/driven-by-big-tech-s-pandemic-gains-s-p-500-s-2020-surge-masks-uneven-recovery-61957736.

2   Challenger, Gray & Christmas. 'Dec 2022 Challenger Report: Job Cuts in 2022 Up 13% Over 2021'. 5 January 2023, https://www.challengergray.com/blog/the-challenger-report-job-cuts-in-2022-up-13-over-2021/.

3   Confino, Paolo. 'Mark Zuckerberg's $46.5 billion loss on the metaverse is so huge it would be a Fortune 100 company—but his net worth is up even more than that'. *Fortune*, 27 October 2023, https://fortune.com/2023/10/27/mark-zuckerberg-net-worth-metaverse-losses-46-billion-earnings-stock/.

4   State of AI Report. https://www.stateof.ai/.

5   State of AI Report. https://www.stateof.ai/.

6   Oberoi, Mohit. 'Why Meta Platforms Stock Still Looks
     Undervalued After Doubling in 2023'. Nasdaq, 6 October 2023,
     https://www.nasdaq.com/articles/why-meta-platforms-stock-still-
     looks-undervalued-after-doubling-in-2023.

7   Financial Times. 'ChatGPT parent OpenAI seeks $86bn valuation'.
     Financial Times, 19 October 2023, https://www.ft.com/content/
     e4ab95c9-5b45-4996-a69e-46075d6428e5.

8   Stone, Brad, and Bergen, Mark. 'OpenAI Is Working With US
     Military on Cybersecurity Tools'. Bloomberg, 16 January 2024,
     https://www.bloomberg.com/news/articles/2024-01-16/openai-
     working-with-us-military-on-cybersecurity-tools-for-veterans.

9   Krantowitz, Alex. 'Oh, Good, OpenAI's Biggest Rival Has a Weird
     Structure Too'. *Slate*, 2 December 2023, https://slate.com/technology/
     2023/12/anthropic-openai-board-trust-effective-altruism.html.

10   Berruti, Massimo. 'Inside the White-Hot Center of A.I.
     Doomerism'. *New York Times*, 12 July 2023, https://www.nytimes.
     com/2023/07/11/technology/anthropic-ai-claude-chatbot.html.

11   Sweney, Mark. 'Amazon to invest up to $4bn in OpenAI rival
     Anthropic'. *The Guardian*, 25 September 2023, https://www.
     theguardian.com/technology/2023/sep/25/amazon-invest-openai-
     rival-anthropic-microsoft-chat-gpt.

12   Crunchbase. 'Current Unicorns Tagged with AI'. https://www.
     crunchbase.com/lists/current-unicorns-tagged-with-ai/f296fc53-ac45-
     44e0-88eb-2979f7857fe2/organization.companies.

13   Mollman, Steve. 'Sam Altman risks sounding "arrogant" to explain
     what's wrong with Silicon Valley—and why OpenAI has no road
     map'. *Fortune*, 7 September 2023, https://fortune.com/2023/09/07/
     sam-altman-silicon-valley-innovation-decline-openai-no-road-map/.

14   Rosoff, Matt. 'Jeff Bezos told what may be the best startup
     investment story ever'. *Business Insider*, 20 October 2016, https://
     www.businessinsider.com/jeff-bezos-on-early-amazon-
     investors-2016-10

15   Comninel, George C. 'English Feudalism and the Origins of
     Capitalism'. *The Journal of Peasant Studies*, Vol. 27 No. 4 (2019),
     pp. 1–53, doi.org/10.1080/03066150008438748.

16   Brenner, Robert. 'Agrarian Class Structure and Economic Development in Pre-Industrial Europe'. *Past & Present*, No. 70 (1976), pp. 30–75.

17   While different scholars' estimates vary, Sherburne Cook puts the figure at around 300,000. See Cook, Sherburne F. *The Population of the California Indians, 1769–1970*. Berkeley: University of California Press, 1976.

18   Madley, Benjamin. *An American Genocide: The United States and the California Indian Catastrophe, 1846–1873*. New Haven: Yale University Press, 2016.

19   Madley, *An American Genocide: The United States and the California Indian Catastrophe, 1846–1873*.

20   Mamdani, Mahmood. *Neither Settler nor Native: The Making and Unmaking of Permanent Minorities*. Cambridge, Massachusetts: The Belknap Press of Harvard University Press, 2020, p. 58.

21   Lightfoot, Kent G., and Parrish, Otis. *California Indians and Their Environment: An Introduction*. Oakland, California: University of California Press, 2009.

22   Coulthard, Glen. *Red Skin, White Masks: Rejecting the Colonial Politics of Recognition*. Minneapolis: University of Minnesota Press, 2014.

23   McWilliams, Carey. *Factories in the Field: The Story of Migratory Farm Labor in California*. Berkeley: University of California Press, 2000.

24   Ngai, Mae M. *Impossible Subjects: Illegal Aliens and the Making of Modern America*. Princeton: Princeton University Press, 2014.

25   Harris, Malcolm. *Palo Alto: A History of California, Capitalism, and the World*. New York: Hachette Book Group, 2023.

26   Stanford was one of the major hubs in the United States for eugenics, a pseudoscientific doctrine of 'improving' the human population through controlled reproduction. This logic had its most extreme iteration in the Nazi Holocaust. Since 2019, the eugenicist history of the university has come under increasing scrutiny, led by the Stanford Eugenics History Project: see www.stanfordeugenics.com.

27   Leslie, Stuart W. *The Cold War and American Science: The*

Military-Industrial-Academic Complex at MIT and Stanford. New York: Columbia University Press, 1993.

28 O'Mara, Margaret. *Cities of Knowledge: Cold War Science and the Search for the Next Silicon Valley*. Princeton: Princeton University Press, 2005.

29 O'Mara, *Cities of Knowledge*.

30 O'Mara, Margaret. *The Code: Silicon Valley and the Remaking of America*. New York: Penguin Press, 2019.

31 O'Mara, *The Code*.

32 O'Mara, *The Code*.

33 Gompers, Paul, and Lerner, Josh. 'The Venture Capital Revolution'. *Journal of Economic Perspectives*, Vol. 15 No. 2 (2001), pp. 145–68, doi.org/10.1257/jep.15.2.145.

34 Gompers and Lerner, 'The Venture Capital Revolution'.

35 Leslie, *The Cold War and American Science*.

36 Barbrook, Richard, and Cameron, Andy. 'The Californian Ideology'. *Mute*, 1 September 1995, https://www.metamute.org/editorial/articles/californian-ideology.

37 Schleifer, Theodore. 'Here are the 15 Silicon Valley millionaires spending the most to beat Donald Trump'. *Vox*, 27 October 2020, https://www.vox.com/recode/21529490/silicon-valley-millionaires-top-donors-2020-election-donald-trump; Schleifer, Theodore. 'Silicon Valley Makes its Anti-Biden Move'. *Puck*, 8 December 2023, https://puck.news/silicon-valley-makes-its-anti-biden-move/.

38 Sherman, Justin. 'Oh Sure, Big Tech Wants Regulation—on Its Own Terms'. *Wired*, 28 January 2020, https://www.wired.com/story/opinion-oh-sure-big-tech-wants-regulationon-its-own-terms/.

39 Stanley, Ben. 'The thin ideology of populism'. *Journal of Political Ideologies*, Vol. 13 No. 1 (2008), pp. 95–110, doi:10.1080/13569310701822289.

40 Jones, Rachyl. 'Mark Zuckerberg's Voting Stake Renders Shareholders Powerless'. *Observer*, 1 June 2023, https://observer.com/2023/06/mark-zuckerberg-2023-shareholder-meeting/.

41 Zuckerberg, Mark. 'Zuckerberg Facebook video live from our weekly internal Q&A' (2019). Zuckerberg Transcripts, 1092,

https://epublications.marquette.edu/zuckerberg_files_
transcripts/1092.

42 For further discussion of the role of meritocratic ideology in
Silicon Valley, see Noble, Safiya Umoja, and Roberts, Sarah T.
'Technological Elites, the Meritocracy, and Postracial Myths in
Silicon Valley'. In *Racism Postrace*, Roopali Mukherjee, Sarah
Banet-Weiser, and Gray Herman (eds.). Duke University Press,
2019, https://doi.org/10.1215/9781478003250.

43 A name for PayPal's founders: a group featuring Tesla's Elon
Musk, YouTube's Steve Chen, and LinkedIn's Reid Hoffman.

44 Thiel, Peter. 'The Education of a Libertarian'. Cato Unbound, 13
April 2009, https://www.cato-unbound.org/2009/04/13/peter-thiel/
education-libertarian/.

45 Slobodian, Quinn. *Globalists: The End of Empire and the Birth of
Neoliberalism*. Cambridge, MA: Harvard University Press, 2018.

46 Iliadis, Andrew, and Acker, Amelia. 'The Seer and the Seen:
Surveying Palantir's Surveillance Platform'. *The Information Society*,
Vol. 38 No. 5 (2022), pp. 334–63, doi.org/10.1080/01972243.2022
.2100851.

7: The Organiser

1 This is not to say that there are no examples of data workers
being able to look to a historical precedent. In 2021, there was a
strike by three unions, the French union Solidaires Unitaires et
Démocratiques (SUD), the Moroccan Labor Union (UMT) and the
Tunisian General Labor Union (UGTT), targeting sites in France,
Morocco and Tunisia simultaneously. Through these strikes,
workers won double-digit pay rises. However, these sorts of
examples are rare and will be unfamiliar to most data workers.

2 On the different 'power resources' workers have access to, see:
Olin Wright, Erik. 'Working-class power, capitalist-class interests,
and class compromise'. *American Journal of Sociology*, Vol. 105
No. 4 (2000), pp. 957–1002, doi/10.1086/210397. See also
Schmalz, Stefan., Ludwig, Carmen, and Webster, Edward. 'The
Power Resources Approach: Developments and Challenges'. *Global*

Labour Journal, Vol. 9 No. 2 (2018), pp. 113–34, mulpress. mcmaster.ca/globallabour/article/view/3569.

3   IWGB. #ShameOnOcado, at: /iwgb.org.uk/en/page/shame-on-ocado/.

4   PAYYOURWORKERS. Open union letter to Adidas CEO Bjørn Gulden, www.payyourworkers.org/open-union-letter-to-adidas-ceo.

5   'Adidas Response to Clean Clothes Campaign Open Letter on PT Panarub Dwikarya Benoa'. https://www.adidas-group.com/media/filer_public/69/1d/691d6520-d1f9-4549-8a94-744dc49ab6ca/adidas_response_to_clean_clothes_campaign_open_letter_on_panarub_dwikarya.pdf.

6   Lipscombe, Dakota and Hirsch, Jeffrey M. 'Google has a union. Is this and other recent organizing efforts a sign of a broader tech labor movement?'. *The North Carolina Journal of Law and Technology*, 8 March 2021, ncjolt.org/blogs/google-has-a-union-is-this-and-other-recent-organizing-efforts-a-sign-of-a-broader-tech-labor-movement/.

7   Hayden Belfield, a researcher at the University of Cambridge, has directly attributed the high bargaining power of AI workers to the limited supply of talent. See Belfield, Hayden. 'Activism by the AI Community: Analysing Recent Achievements and Future Prospects', *Proceedings of the AAAI/ACM Conference on AI, Ethics, and Society* (2020), pp. 15–21, doi/10.1145/3375627.3375814. Corroborating this, a 2022 OECD report has shown 33 per cent growth in vacancies requiring AI skills in Australia, Canada, NZ, UK, US, Austria, Belgium, France, Germany, Italy, Netherlands, Spain, Sweden and Switzerland between 2019 and 2022. See Borgonovi, Francesca, et al. 'Emerging trends in AI skill demand across 14 OECD countries', www.oecd-ilibrary.org/docserver/7c691b9a-en.pdf. Looking at the technology sector more broadly, a 2023 McKinsey report has shown that job postings grew 15 per cent between 2021 and 2022, despite global job postings decreasing by 13 per cent overall. A survey of 3.5 million job postings across fifteen tech trends found that many of the skills in greatest demand have less than half as many qualified practitioners per posting as the global average. See Chui, Michael, et al. 'McKinsey Technology

Trends Outlook 2023', www.mckinsey.com/capabilities/mckinsey-digital/our-insights/the-top-trends-in-tech#tech-trends-2023.

8   In the US, for instance, the median tech occupation wage (including tech professionals and business professionals employed at tech companies) in 2021 (the latest available government data) was $100,615. This is 103 per cent higher than the median wage for the entire nation. See: 'The Computing Technology Industry Association, State of the Tech Workforce', www.cyberstates.org/pdf/CompTIA_State_of_the_tech_workforce_2023.pdf (p. 8).

9   Maxwell, Thomas. 'Tech is paying big salaries for engineering and developer jobs specializing in AI. Here's how much you can make'. Business Insider, n.d, www.businessinsider.com/ai-engineer-developer-salary-jobs-2023-6?r=US&IR=T.

10   Tarnoff, Ben. 'The Making of the Tech Worker Movement'. *Logic(S) Magazine*, 9 May 2020, logicmag.io/the-making-of-the-tech-worker-movement/full-text/.

11   Neveragain.tech. 'Our Pledge', neveragain.tech.

12   Tarnoff, 'The Making of the Tech Worker Movement'.

13   Tarnoff, 'The Making of the Tech Worker Movement'.

14   Biddle, Sam. 'Of Nine Tech Companies, Only Twitter Says it Would Refuse to Help Build Muslim Registry for Trump'. *The Intercept*, 2 December 2016, theintercept.com/2016/12/02/of-8-tech-companies-only-twitter-says-it-would-refuse-to-help-build-muslim-registry-for-trump/.

15   Tech Workers Coalition. 'Tech Workers, Platform Workers, and Workers' Inquiry'. *Notes From Below*, 30 March 2018, notesfrombelow.org/article/tech-workers-platform-workers-and-workers-inquiry.

16   Collective Action in Tech. data.collectiveaction.tech.

17   Smith, Brad. 'Microsoft Adopts Principles for Employee Organizing and Engagement with Labor Organizations'. *Microsoft On the Issues*, 2 June 2022, blogs.microsoft.com/on-the-issues/2022/06/02/employee-organizing-engagement-labor-economy/; Silberling, Amanda. 'Microsoft Now Has Its First Official Union in the US'. TechCrunch, 3 January 2023, techcrunch.com/2023/01/03/microsoft-zenimax-union/.

18   Wong, Julia C, and Koran, M. 'Google Contract Workers in

Pittsburgh Vote to Form Union'. *The Guardian*, 25 September 2019, www.theguardian.com/technology/2019/sep/24/google-contract-workers-in-pittsburgh-vote-to-form-union; Kari, Paul. 'Kickstarter Workers Vote to Unionize amid Growing Industry Unrest'. *The Guardian*, 19 February 2020, www.theguardian.com/us-news/2020/feb/18/kickstarter-union-crowdfunding-tech.

19 Collective Action in Tech.

20 Tech Workers Coalition. 'Tech Workers, Platform Workers, and Workers' Inquiry', 2018.

21 Klint, Finley. 'How GitHub Is Helping Overworked Chinese Programmers'. *Wired*, 4 April 2019, www.wired.com/story/how-github-helping-overworked-chinese-programmers/.

22 Klint, 'How GitHub Is Helping Overworked Chinese Programmers'.

23 MSWorkers, Microsoft and GitHub Workers Support 996.ICU, github.com/MSWorkers/support.996.ICU.

24 Von Struensee, Susan. 'The Role of Social Movements, Coalitions, and Workers in Resisting Harmful Artificial Intelligence and Contributing to the Development of Responsible AI'. 16 June 2021, https://ssrn.com/abstract=3880779, p. 22. Google figures come from March 2019, see: Wakabayashi, Daisuke. 'Google's Shadow Work Force: Temps Who Outnumber Full-Time Employees'. *New York Times*, 28 May 2019, www.nytimes.com/2019/05/28/technology/google-temp-workers.html.

25 Stone, Katherine. 'The Origins of Job Structures in the Steel Industry'. *Review of Radical Political Economics*, Vol. 6 No. 2 (1974), pp. 113–73, doi.org/10.1177/048661347400600207.

26 Tarnoff, 'The Making of the Tech Worker Movement'.

27 Tarnoff, 'The Making of the Tech Worker Movement'.

28 Stapleton, Claire, et al. 'We're the Organizers of the Google Walkout. Here Are Our Demands'. *The Cut*, 1 November 2018, www.thecut.com/2018/11/google-walkout-organizers-explain-demands.html.

29 Bhuiyan, Johana. 'The Google walkout: What protesters demanded and what they got'. *Los Angeles Times*, 6 November 2019, www.latimes.com/business/technology/story/2019-11-06/google-walkout-demands.

30 Following the corporate restructuring of Google in the Alphabet conglomerate in 2015, Alphabet replaced 'Don't be evil' with 'Do the right thing'.

31 Tarnoff, Ben. 'Tech Workers Versus the Pentagon: Interview with Kim'. Jacobin, 6 June 2018, jacobin.com/2018/06/google-project-maven-military-tech-workers.

32 Tarnoff, Ben, 'Tech Workers Versus the Pentagon'.

33 Von Struensee, Susan, 'The Role of Social Movements, Coalitions, and Workers in Resisting Harmful Artificial Intelligence', 16 June 2021, pp. 21–2.

34 Collective Action In Tech. https://data.collectiveaction.tech/action/117.

35 JEDI was ultimately terminated due to conflict between Amazon and Microsoft over procurement. Its successor, JWCC, involved work from Amazon, Microsoft, Oracle and Google.

36 Collective Action In Tech, https://data.collectiveaction.tech/action/568.

37 Collective Action In Tech. https://data.collectiveaction.tech/action/569; 'No Tech for Apartheid: Protesters Call Out Google for Cloud Contract with Israel'. DemocracyNow!, 1 September 2023, www.democracynow.org/2023/9/1/headlines/no_tech_for_apartheid_protesters_call_out_google_for_cloud_contract_with_israel.

38 Although in the case of the anti-Maven protests, it is important to note that organising efforts also crucially relied on one key team of workers who refused to work. See Molinari, Carmen. 'There is something missing from tech worker organizing'. Organizing Work, 9 December 2020, organizing.work/2020/12/there-is-something-missing-from-tech-worker-organizing/.

39 After 2021, there was a shift in the type of activity that tech workers were engaging in: from petitions and public statements towards unionisation. See: Lytvynenko, Jane. 'Why the balance of power in tech is shifting toward workers'. MIT Technology Review, 7 February 2022, www.technologyreview.com/2022/02/07/1044760/tech-workers-unionizing-power/.

40 Clean Clothes Campaign. cleanclothes.org.

41 Clean Clothes Campaign. The Accord on Safety, cleanclothes.org/campaigns/the-accord.

42 Clean Clothes Campaign, The Accord on Safety.

43 ExChains. www.exchains.org.

44 Lohmeyer, Nora., Schüßler, Elke, Helfen, Markus. 'Can solidarity be organized "from below" in global supply chains? The case of ExChains'. *Industrielle Beziehungen*, Vol. 25 No. 4 (2018), pp. 401–24. www.econstor.eu/bitstream/10419/227425/1/indbez-v25i4p400-424.pdf (pp. 5–6).

45 Kuttner, Robert. 'Labor's Militant Creativity'. *The American Prospect*, 18 September 2023, prospect.org/blogs-and-newsletters/tap/2023-09-18-labors-militant-creativity/.

46 Fricke, Beatrix. 'BSR: Streik auch am Freitag – Verzögerungen bei Müllabfuh'. *Logo Berliner Morgenpost*, 10 March 2023, www.morgenpost.de/berlin/article237845375/streik-bsr-berlin-muellabfuhr.html.

47 Kassem, Sarrah. 'Labour realities at Amazon and COVID-19: obstacles and collective possibilities for its warehouse workers and MTurk workers'. *Global Political Economy*, Vol. 1 No. 1 (2022), pp. 59–79. bristoluniversitypressdigital-com.ezproxy-prd.bodleian.ox.ac.uk/view/journals/gpe/1/1/article-p59.xml.

8: Rewiring the Machine

1 Bloch, Ernst. *The Principle of Hope*. Cambridge, MA: The MIT Press, 1995.

2 Ober, Josiah. *Mass and Elite in Democratic Athens: Rhetoric, Ideology, and the Power of the People*. Princeton, New Jersey: Princeton University Press, 1989.

3 Cooper, Frederick. *Decolonization and African Society: The Labor Question in French and British Africa*. Cambridge: Cambridge University Press, 1996.

4 Ransbury, Barbara. *Ella Baker and the Black Freedom Movement: A Radical Democratic Vision*. Chapel Hill: The University of North Carolina Press, 2005.

5 Douglass, Frederick. 'West India Emancipation, speech delivered at Canandaigua, New York, 3 August 1857', rbscp.lib.rochester.edu/4398.

6 O'Neill, Martin, and White, Stuart. 'Trade Unions and Political

Equality'. In: Hugh Collins, Gillian Lester, Virginia Mantouvalou (eds.), *Philosophical Foundations of Labour Law*. Oxford: Oxford University Press, 2018.

7   Klein, Steven. 'Democracy Requires Organized Collective Power'. *Journal of Political Philosophy*, Vol. 30 No. 1(2021), pp. 26–47, https://onlinelibrary.wiley.com/doi/full/10.1111/jopp.12249.

8   McGaughey, Ewan. 'The Codetermination Bargains: The History of German Corporate and Labour Law'. *Columbia Journal of European Law*, Vol. 23 No. 1 (2016), pp. 1–43 (p. 35). https://ssrn.com/abstract=2579932.

9   Jäger, Simon, Noy, Shakked and Schoefer, Benjamin. 'What Does Codetermination do?'. NBER Working Paper Series (2016), https://www.nber.org/system/files/working_papers/w28921/w28921.pdf.

10 Ferreras, Isabelle. *Firms as Political Entities: Saving Democracy through Economic Bicameralism*. Cambridge: Cambridge University Press, 2017.

11 Viktorsson, Mio Tastas and Gowan, Saoirse. 'Revisiting the Meidner Plan'. Jacobin, 22 August 2017, https://jacobin.com/2017/08/sweden-social-democracy-meidner-plan-capital.

12 Lawrence, Mat and Mason, Nigel. 'Capital gains: Broadening company ownership in the UK economy'. Institute for Public Policy Research, 21 December 2017, www.ippr.org/research/publications/CEJ-capital-gains.

13 Syal, Rajeev. 'Employees to be handed stake in firms under Labour plan'. *The Guardian*, 24 September 2023, www.theguardian.com/politics/2018/sep/23/labour-private-sector-employee-ownership-plan-john-mcdonnell.

14 The Fairwork AI principles stipulate that workers should be paid at least the local living wage, that they should be protected from foundational risks that arise from doing their jobs, that contracts should be accessible and fair, that equity and due process should be ensured in the way workers are managed, and that all workers have channels for collective voice and collective representation. See: Fairwork, Principles, https://fair.work/en/fw/principles/.

15 Graham, Mark, et al. 'The Fairwork Foundation: Strategies for

Improving Platform Work'. In: *The Weizenbaum Conference 2019 Challenges of Digital Inequality – Digital Education, Digital Work, Digital Life*, pp. 1–8, doi.org/10.34669/wi.cp/2.13.

16 Fairwork 2023 Annual Report. Oxford.

17 Sama. 'The Ethical AI Supply Chain: Purpose-Built for Impact', https://www.sama.com/ethical-ai/.

18 GPAI. Fairwork AI Ratings 2023: The Workers Behind AI at Sama. Report, December 2023, Global Partnership on AI, https://gpai.ai/ projects/future-of-work/FoW-Fairwork-AI-Ratings-2023.pdf.

19 GPAI. Fairwork AI Ratings 2023: The Workers Behind AI at Sama.

20 Google. Supplier Responsibility Report 2022, www.gstatic.com/ gumdrop/sustainability/google-2022-supplier-responsibility-report. pdf.

21 For information on the bills, see: Republic of Philippines, House of Representatives, House Bills and Resolutions, https://www.congress. gov.ph/legisdocs/.

22 Robinson, J. *Economic Philosophy*. Piscataway, NJ: Aldine, 2006 [first published 1962], p. 45.

23 'Italy: Amazon reaches agreement with Italian unions', *Industrial Relations and Labour Law Newsletter*, November 2021, https:// industrialrelationsnews.ioe-emp.org/industrial-relations-and-labour-law-november-2021/news/article/italy-amazon-reaches-agreement-with-italian-unions.

24 Once adopted in 2024, it is expected that its provisions will be applicable to companies from 2027 onwards.

25 See, for instance, this proposed ILO convention on platform work: Fredman, S., et al. 'International Regulation of Platform Labor: A Proposal for Action'. *Weizenbaum Journal of the Digital Society*, Vol. 1 No. 1 (2021), https://doi.org/10.34669/wi.wjds/1.1.4.

26 Cooperatives Europe. 'The Power of Cooperation: Cooperative Europe's Key Figures 2015'. April 2016, https://coopseurope.coop/ wp-content/uploads/files/The%20power%20of%20Cooperation%20 -%20Cooperatives%20Europe%20key%20statistics%202015.pdf.

27 An, Jaehyung., Cho, Soo-Haeng and Tang, Christopher S. 'Aggregating Smallholder Farmers in Emerging Economies. Production and Operations Management'. *Production and Operations*

Management, Vol. 29 No. 9 (2015), pp. 1,414–29, doi.org/10.1111/poms.12372.

28   International Cooperative Alliance. Celebrating 125 Years, www.ica.coop/en/celebrating-125-years.

29   Karya. What we do, karya.in/about/work.html.

30   Miller, Katherine. 'Radical Proposal: Data Cooperatives Could Give Us More Power Over Our Data'. Stanford University: Human Centered Artificial Intelligence, https://hai.stanford.edu/news/radical-proposal-data-cooperatives-could-give-us-more-power-over-our-data.

31   Scholz, Trebor. *Own This!: How Platform Cooperatives Help Workers Build a Democratic Internet*. London: Verso, 2023.

Conclusion

1   Ahronheim, Anna. 'Israel's operation against Hamas was the world's first AI war'. *Jerusalem Post*, 27 May 2021, https://www.jpost.com/arab-israeli-conflict/gaza-news/guardian-of-the-walls-the-first-ai-war-669371.

2   Interview originally given to YNet. Quoted in *The Guardian*. https://www.ynet.co.il/news/article/byatqqx00h. Davies, Harry, McKernan, Bethan, Sabbagh, Dan. '"The Gospel": How Israel uses AI to select bombing targets in Gaza. *The Guardian*, 1 December 2023, https://www.theguardian.com/world/2023/dec/01/the-gospel-how-israel-uses-ai-to-select-bombing-targets.

3   Fisk, Robert. *Pity the Nation: Lebanon at War*. Oxford: Oxford University Press, 1990.

4   Browne, Malcolm W. 'Invention That Shaped the Gulf War: the Laser-Guided Bomb'. *New York Times*, 26 February 1991, https://www.nytimes.com/1991/02/26/science/invention-that-shaped-the-gulf-war-the-laser-guided-bomb.html.

5   Abraham, Yuval. '"A mass assassination factory": Inside Israel's calculated bombing of Gaza'. +972 *Magazine*, 30 November 2023, https://www.972mag.com/mass-assassination-factory-israel-calculated-bombing-gaza/.

6   Abraham, '"A mass assassination factory"'.

7   McKernan, Bethan, and Kierszenbaum, Quique. '"We're focused on

maximum damage": ground offensive into Gaza seems imminent'. *The Guardian*, 10 October 2023, https://www.theguardian.com/world/2023/oct/10/right-now-it-is-one-day-at-a-time-life-on-israels-frontline-with-gaza.

8   Abraham, '"A mass assassination factory"'.

9   Abraham, Yuval. '"Lavender": The AI machine directing Israel's bombing spree in Gaza'. +972 *Magazine*, 3 April 2024, https://www.972mag.com/lavender-ai-israeli-army-gaza.

10   Loewenstein, Antony. *The Palestine Laboratory: How Israel Exports the Technology of Occupation around the World*. London: Verso, 2023.

11   Biddle, Sam. 'Documents Reveal Advanced AI Tools Google Is Selling to Israel'. The Intercept, 24 July 2022, https://theintercept.com/2022/07/24/google-israel-artificial-intelligence-project-nimbus/.

12   Jewish Diaspora in Tech. 'Letter on Google Communication and Cloud Contracts'. https://docs.google.com/forms/d/e/1FAIpQLSck0yb4OgXJlf5N2Ap21BhW-QNTL0rigFSYsW_o09QkJOJVbw/viewform.

13   Detrixhe, John. 'The tech industry is hiring Israeli engineers as fast as the army can produce them'. *Quartz*, 15 August 2017, https://qz.com/1050844/the-tech-industry-is-hiring-israeli-engineers-as-fast-as-the-army-can-produce-them.

14   Mohamed, Shakir, Png, Marie-Theres, and Isaac, William. 'Decolonial AI: Decolonial Theory as Sociotechnical Foresight in Artificial Intelligence'. *Philosophy & Technology*, Vol. 33 (2020), pp. 659–84; Wallerstein, Immanuel. *The Modern World-System: Capitalist Agriculture and the Emergence of the European World Economy in the Sixteenth Century*. New York: Academic Press, 1976; Grosfoguel, Ramón. 'Developmentalism, Modernity and Dependency Theory in Latin America'. In Mignolo, Walter (ed.). *Nepantla: Views from South*, Durham: Duke University Press, 2000.

15   EuroNews. 'The future of work lies in the balance between human and AI'. EuroNews, 7 December 2023, https://www.euronews.com/my-europe/2023/12/06/the-future-of-work-lies-in-the-balance-between-human-and-ai.

16   Hern, Alex. 'Microsoft productivity score feature criticised as workplace surveillance'. *The Guardian*, 26 November 2023, https://

www.theguardian.com/technology/2020/nov/26/microsoft-productivity-score-feature-criticised-workplace-surveillance.

17   Kak, Amba, and West, Sarah Myers. 'AI Now 2023 Landscape: Confronting Tech Power'. AI Now Institute, 11 April 2023, https://ainowinstitute.org/2023-landscape.

18   CWU. 'AI the "Most Significant Threat" to Workers Today, CWU Urges'. CWU, 2023, https://www.cwu.org/news/ai-the-most-significant-threat-to-workers-today-cwu-urges/.

19   Moody, Kim. *On New Terrain: How Capital Is Reshaping the Battleground of Class War*. Chicago, Illinois: Haymarket Books, 2017, p. 50.

20   The speech can be found on YouTube, including at: Indrid Cold, 'Mario Savio | Bodies Upon The Gears'. https://www.youtubåe.com/watch?v=xz7KLSOJaTE&ab_channel=IndridCold.